The History of the
ABACUS

The History of the
ABACUS

J. M. Pullan

PRAEGER PUBLISHERS
New York · Washington

BOOKS THAT MATTER

Published in the United States of America in 1969
by Praeger, Publishers Inc.,
111 Fourth Avenue, New York, N.Y. 10003

Second printing 1970

Library of Congress Catalog Card Number: 72–75113

Printed in Great Britain

ACKNOWLEDGMENTS

It is more than thirty years since the author's interest in the teaching of arithmetic to young children was aroused in the schools of Manchester and Salford. He remembers with pleasure many discussions over the years, with teachers and colleagues in the Inspectorate, on ways of dealing with this most troublesome of the three Rs. He is glad to acknowledge the inspiration gained many years ago from Mr L. C. Schiller, later Staff Inspector for Primary Education, and, more recently, valuable help in the correction of the manuscript given by a former colleague, Miss W. E. Deavin. He is especially grateful to Dr Michael Lewis and Mrs Hilda Lewis for their interest and encouragement and for their advice on educational and historical matters, and to Miss E. M. Horsley, the publisher's editor, for her understanding and guidance during the process of converting a rather rough manuscript into a published book.

Much of the information used in the book was gathered in visits to numerous museums and libraries, or in correspondence with them, and the author wishes to express his gratitude to members of their staffs for their patience and readiness to give assistance and advice. The help given in various departments of the British Museum, in the Heberden Coin Room of the Ashmolean Museum, and on many occasions in the City Museum, Bristol, was especially valuable.

Permission to use photographs was given by the British Museum (the Babylonian tablet, Egyptian papyrus, Roman abacus and the extract from Recorde's *Ground of Artes*, pp. 3, 10, 19, and 49 respectively), the City Museum Bristol (Roman calculi, p. 21), the Cabinet des Médailles, Bibliothèque

Nationale, Paris (Roman calculator, p. 27), the Bristol City Archivist (Wine Street rents, pp. 40 and 41), the Public Record Office (the Treasury documents, pp. 44, 45, 46 and 47), and the Musées de la Ville de Strasbourg (the 'table de payeurs', p. 53). Special thanks are due to a Mexican scholar, Fratelli Luis Benavides, whose thesis *Pedagogia del Abaco* describes the bead-frame type of abacus, for permission to use the illustrations of the sculpture of a dying Roman (p. 29) and the Aztec vase (p. 93), and the author was delighted to have an opportunity to discuss the history of the abacus with him. The photographs on pp. 59 and 60 are reproduced from books in the private library of the late Professor F. P. Barnard by permission of the Ashmolean Museum.

CONTENTS

Introduction   *page* ix

I The origins of arithmetic   1

II Calculus and abacus   16

III Roman numerals to Arabic figures   30

IV Counter-casting to pen-reckoning   43

V The abacus method   57

VI Jettons   71

VII The abacus in archaeology   89

VIII The abacus in education   94

Conclusion   102

Appendix   105

Pebbles or counters used to build up a table   105

Finding square roots without calculation   106

Calculating with counters   107

Literary references to the use of counters   110

Classical references to the use of the abacus   113

Greek abaci   114

Makers of Nuremberg tokens   115

Bibliography   116

Notes   121

Index   125

INTRODUCTION

'Let me see. Every 'leven wether tods; every tod yields pound and odd shilling; fifteen hundred shorn, what comes the wool to? . . . I cannot do't without counters.'

A WINTER'S TALE, iv, 2

In Shakespeare's time the old method of reckoning with counters on a board was still the only way in which most people could deal with calculations too difficult to manage in their heads. The clown in *A Winter's Tale* had set himself a calculation that was difficult, but not at all unusual among shepherds at shearing-time. It might be expressed in modern textbook language: 'How much money can be obtained for the wool from 1500 sheep (wethers) if every eleven sheep yield one tod (28 lb.), each tod being worth a guinea?'[1] No wonder he needed counters!

The use of written Arabic figures was known only to a few scholars in those days, and even 'a great arithmetician, one Michael Cassio', in the opening scene of *Othello*, was described as 'this counter-caster'. Numbers were still usually written in the Roman style.

It is little more than two hundred years since 'pen-reckoning' came into general use in the everyday affairs of merchants, bankers and accountants, replacing the older method of 'counter-casting'. Although Arabic figures had been known in Western Europe for several centuries they took the place of Roman numerals very gradually, and in Shakespeare's time they were still not understood by ordinary folk. The way of using them remained a mystery to the great majority of people, who could neither read nor write, until towards the end of the seventeenth century, and even later in some countries.

The traditional method of reckoning with counters was efficient enough, well understood and trusted, for it had been in use, with little variation, for hundreds, or even thousands, of years. Where numbers or amounts of money were small enough they could be dealt with mentally, and much simple reckoning could be effected in a practical manner. For example, the difference in length between two ropes could be found simply by laying them side by side; and money could be added by placing the actual coins on a table. Even at an early stage in the development of civilisation, however, occasions often arose when these methods were not enough; but by the use of counters, and with an understanding of 'place value', the more difficult computations could easily be managed.

The practice of using brass counters, often called 'jettons',[2] is said to have been brought to this country by the Normans soon after the Conquest; but it was not new. It was, in fact, an up-to-date version of a method commonly used in Greece and Rome more than a thousand years before. Instead of metal counters the Greeks and Romans had used pebbles or discs of glass, bone or ivory called 'pessoi' or 'calculi', and their table was an 'abacus'.[3] The origin of the method is unknown but it was used in Greece five hundred years B.C. and may have come from the earlier civilisations of Asia Minor, Mesopotamia or India.

When Arabic figures came into general use in account books and other records, early in the seventeenth century, the old method of reckoning with counters continued for a time; and even a century later it was still recommended for those who could not read or write. Pen-reckoning with Arabic figures depended upon the same principle, place value, as the earlier method; but it seems that this important fact was not generally recognised.

One of the earliest European mathematicians to understand and use Arabic figures, Gebert, who became Pope Sylvester II in the year 999, arranged his written figures carefully in columns and treated them as counters had been treated on an abacus. But later some mathematicians, and many unlearned people who tried to use the new system, allowed themselves to be confused by the mysterious zero, which had no parallel in the Roman system.

Arabic notation had two great advantages: it could be written more neatly than Roman, especially when large numbers were concerned, and it had unlimited scope. When, in the eighteenth and nineteenth centuries, industry and commerce began to develop rapidly, the keeping of accounts grew in complexity and in the size of the numbers involved; and the saving of time began to matter. Neither the Roman system of notation nor the counting-board could any longer meet the growing needs of the times.

Schoolmasters and the writers of textbooks now came into the picture. Many of them were concerned with the basic needs of an educated working class, especially the black-coated office clerks, and they took to the teaching of written arithmetic with enthusiasm. Not only was it a subject that could be directed towards the needs of business affairs, but it was an excellent form of mental discipline. There was also the question of physical discipline, for classes were large and conditions poor. 'Sums' could be used to keep a large class quiet!

Teaching became more and more formal, with masses of 'mechanical' exercises of little practical value; and it is only in recent years that schoolmasters have thrown off the deadening effect of a hundred years of a blind '3R' tradition in arithmetic. Children are now encouraged to understand the principles involved and to see how they are used outside the classroom. Nevertheless, emphasis in schools is still upon hand-written figures, in spite of the growing use in real life of mechanical devices from cash-registers to electronic computers.

A study of the older methods of manual computing will help to show that all three methods, counter-casting, pen-reckoning and modern mechanical calculating, are based upon the same simple principles.

As a starting-point we can note the survival in modern languages of several words that originated in these older methods. 'Counter', 'check' or 'cheque', and especially the word 'calculate', were all used long before pen and paper had any place in everyday arithmetic. 'Calculate' takes us back at once to Roman times when 'calculi' were in general use for reckoning; and the word 'abacus', which meant the board or table on which the calculi were used, and which is still used to denote any kind of calculating device, comes from an ancient word of Semitic origin.

It would seem desirable, therefore, to go right back to the earliest stages in the development of arithmetical ideas, examining ways in which basic ideas of counting and reckoning may have arisen and following them through successive historical periods.

For the later stages there is abundant evidence in the form of written records, illustrations and a large variety of the actual brass counters used, together with a few remaining tables and reckoning-cloths; but, inevitably, the further back one goes the less hope there can be of finding firm evidence upon which to base conclusions. A certain amount of guess-work, therefore, is unavoidable, especially about the possible use of the abacus method in prehistoric times; but there is enough evidence from archaeological sources to justify the belief that some such method of reckoning preceded the evolution of written notations, and that the basic principles have persisted, varying only in form and appearance, from the earliest times to the present day.

In Chapter One the way in which the first ideas of counting, and then of reckoning, may have arisen are considered. Chapter Two deals with archaeological evidence, supported by literary references, of the early use of 'pebbles', especially calculi, and the simple flat abacus. Doubts about the origin of the word abacus, and reasons for the modern use of the word in relation to other calculating devices, especially the old Roman bead-frames and modern Asiatic versions, are explored.

The widespread use in the West, over a long period of time, of the apparently clumsy Roman notation, and its gradual replacement by Arabic, need careful consideration. Chapter Four is devoted to the origins of the Roman system and its relation to the use of the abacus and, later, of the counter-board and the Exchequer Table. Evidence from documents and text-books shows that the use of counters continued for some time after the general change, in the first half of the seventeenth century, to the new figures. Pen-reckoning using Arabic notation did not come easily to the merchants and accountants before the spread of popular education and the older method of casting accounts did not disappear until the latter part of the eighteenth century.

In Chapter Five various forms of the abacus method are described, with examples worked out in detail. Attention is given to the way in which the Exchequer Table was used from the establishment of the English Exchequer in the twelfth or thirteenth century until the tables disappeared about four hundred years later.

The jettons, or brass counters, are not only interesting in themselves, but they often show evidence, in their designs or inscriptions, of the ways in which they were used. Some are mentioned earlier, but a general account of them is given in Chapter Six with a few typical examples, mainly from the author's small private collection.

Evidence concerning the use of the abacus in classical times is abundant enough to show that the ability of people of those days to deal with arithmetical problems was higher than is sometimes supposed; and there are reasons for thinking that simple methods of reckoning with counters may have been known in very much earlier times. These matters, and some of their implications, are discussed in Chapter Seven.

At the other end of the time scale we have the present-day problem of finding suitable ways of introducing young children to the intricacies of Arabic arithmetic and, later, of explaining to them the principles of modern mechanical computation. It is suggested in Chapter Eight that a study of the history of the abacus method would be of benefit to teachers and might be of considerable interest to the children.

I

THE ORIGINS OF ARITHMETIC

'What would life be without arithmetic . . . ?'
THE REV. SYDNEY SMITH, 1835

'The knowledge of the origin of the invention of Arithmetic is lost in the unwritten annals of the past, and is engulphed in the whirlpool of unpiercable obscurity.' With this observation W. S. R. Waddrington, Esq., F.R.A.S., proceeded to write, in 1842, a textbook on *Mechanical Arithmetic* which began with a definition of 'Practical Arithmetic' as the art of computing with the ten Arabic figures, and continued by way of the simple and compound rules until he had covered all the exercises and problems that were needed for the banking, insurance and other commercial transactions of the day. They included 'The Rule of Five' as well as 'The Rule of Three', 'Involution and Evolution', 'Fellowship','Medial and Alternate Alligation', and, not surprisingly, a chapter of 'Promiscuous Questions'.

Waddrington was one of the nineteenth-century teachers who were enthusiastic about the 'illimitable use and the boundless application' of the 'famous science of Arabic Arithmetic', still a comparatively new way of calculating as far as the majority of ordinary people were concerned. It had long been known to mathematicians, but in the leisurely days before the rapid development of industry and commerce during the eighteenth and nineteenth centuries the traditional method, counter-casting, had sufficed to deal with computations necessary for normal everyday business transactions. Some merchants still had their 'counting-boards', and taxes were still reckoned on the Exchequer Table in the eighteenth century. Brass counters (*rechen-pfennigs*) were still being manufactured in large numbers in Nuremberg for use in England and other countries during the reign of George II. In France 'jetons' were still in common use for reckoning up to the time of Louis XV. Some counting-houses throughout Europe clung to the old method even later than this, and the British Exchequer was not entirely free of it until 1826.[1]

Early books on arithmetic, from the sixteenth century onwards, taught reckoning with counters as well as with the pen; and Arabic arithmetic spread only slowly. Even those who could read the books, a small minority, often preferred the safer, familiar method of reckoning, using Arabic notation at first only for recording purposes. It was not until the spread of popular education, much later, that tradesmen and their customers began to use Arabic figures to reckon with.

The methods used for reckoning with counters varied a little, and at first sight there would seem to be a considerable difference between the operations carried out on the exchequer table and the calculations with simple numbers on the counter-board; but all counter-casting followed the basic principles of the abacus.

The normal abacus of the Greeks and Romans was a plain board or table on which a few parallel lines were drawn to mark the 'places'. The method was known in Greece five hundred years or more B.C., but was certainly much older than this. Its origin is obscure but it may well have been devised at a very early stage in the development of mathematical ideas.

A great deal is now known about ancient civilisations such as those of Egypt, Mesopotamia, India and China, in which mathematical thought and practices were well advanced long before they appeared in Europe. Records written on papyrus, inscribed on clay tablets or painted on walls show how men were able to measure and use numbers in ways that necessitated some form of calculation several thousand years ago. A remarkable example is a clay tablet unearthed at Senkereh, near Babylon, in 1854, and now in the British Museum.

The tablet shows a list of the squares of numbers up to twenty-four and beyond, and was used, presumably, as a 'ready-reckoner'. Consecutive numbers are shown in the middle column with their squares to the left. The cuneiform

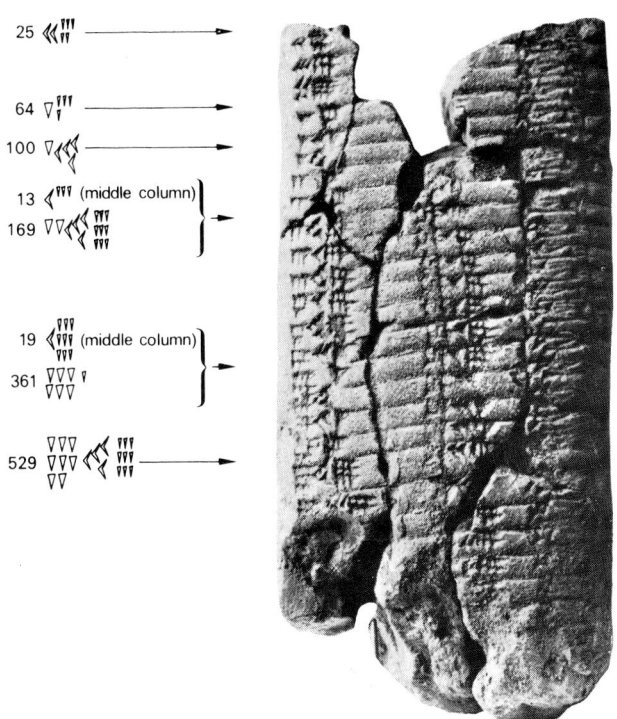

The Senkereh Tablet, 2300–1600 B.C.

notation being a three-dimensional 'script', none of the conventional printed forms is really accurate, and a much better idea of how they looked is given by a photograph of the numbers pressed on plasticine (*see* p. 4), where the impressions in the top row (left to right) represent one, ten and (when this sign appears to the left of other figures) sixty; in the second row, seventy-one; in the third, twenty-five; and in the bottom row, 529. One tends to assume that a special wedge-shaped tool would be needed to produce wedge-shaped marks, but all that is actually required is a piece of wood, etc., with an angled corner, as shown to the left of the photograph. Nothing is known of the way in which the calculations were made, but the tablet is proof that some method of multiplying

B

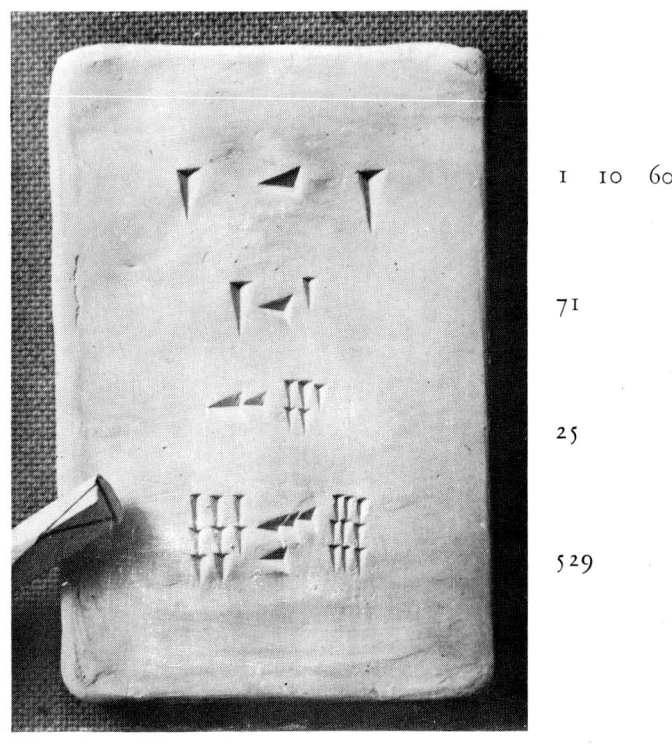

I	10	60
		71
		25
		529

Specimen cuneiform impressions on plasticine and the piece of wood with the angled corner (left) used to make them

numbers was known in Babylon about two thousand years B.C.

Long before this time the earliest primitive ideas of number must have arisen from ordinary practical situations in which there was need to communicate, by gesture or speech, information about how many, how big, how far and how long. Simple counting and elementary reckoning with, at first, only small numbers would be accompanied by movements of arms or fingers, perhaps with some kind of spoken sound, and later by the use of sticks, stones or other objects as symbols.

These ideas and practices would not arise suddenly, of course, or at any one time in the history of mankind. They would probably appear at different times and in different places, coming and going spasmodically at any time during the tens of thousands of years that we now call the 'Stone Age'; and certainly long before, and therefore quite independent of, any kind of written language.

At some later stage, but still a matter of thousands of years B.C., and long before the Babylonian tablets or the invention of any system of 'written' notation, there must have come an important break-through in the discovery of the concept of a 'group number'. This may, perhaps, have occurred in the course of oral counting accompanied by the showing of fingers. There would be a natural pause at five, with a longer break at ten. With one hand five units make a group; if both hands are used the group becomes one of ten units. It was a step forward of the greatest significance when men began to think in these groups and to count them, as well as the units of which they are composed, using the same number language.

How important this step was in the extension of number concepts from 'the fingers of one hand' to ability to think of much larger numbers in precise terms may be realised by remembering that there are still primitive native tribes in existence whose number language does not extend beyond 'one, two, three'.

Ten, rather than five or any other possible basic group number, has been used almost universally from the earliest times. In some ways it was not a good choice because it has only two factors; twelve would have been better from a mathematical point of view, being divisible by two, three, four and six. However, the use of ten as the normal 'base' became established through, no doubt, the convenience of counting with fingers, long before men capable of considering the mathematical implications appeared on the scene.

The limitations of ten are shown by the use of various other group numbers to meet the special requirements of practical situations. Few ancient systems of money, weights or measures made much use of ten as a base, more convenient group numbers being chosen as the need arose; but for counting and reckoning there has rarely been any attempt to use any other base than ten.

Whatever group numbers may have been used, however, there must have been some early stage when there would be well-established oral counting systems in which the units of the basic group had individual names; and it would follow that the groups could be counted using the same words, just as we might say, in our own language, 'four tens and four units'. But in representing numbers by using concrete symbols such as pebbles instead of sounds, a new feature arose, the concept of 'place value'.

This might have arisen in a very simple way from the everyday needs of simple people, herdsmen, farmers or builders. A shepherd, for example, might put pebbles in a heap as he counted his sheep, and he might have hit upon what is really quite a simple device but one which would enable him to count a large number of sheep using only a few pebbles. This would be to count a small fixed number, say ten, over and over again, putting a pebble in a second 'place' each time he completed the 'set'. For example, twenty-seven would appear like this:

Number of groups,
each of, say, ten units.

Units

This would come as a natural accompaniment to rhythmical oral counting, associated with rhythmical hand movements.

In some such way the first use of the principle of place value could have arisen. Knowledge of this way of using counters, and skill in practising it, would spread slowly and unevenly, following the gradual evolution of social groups living by hunting and farming, and gradually acquiring the arts of weaving, the making of pottery, working in metal and building. As communities became more highly organised and trade developed, there might arise, at times, the need for something more than mere counting, or measuring and working by 'rule of thumb'; and it is conceivable that the use of pebbles, counted in groups, was followed, even at a very early stage, by an elementary form of calculation on the principle of what was later called the abacus.

Once the custom was established of counting up to a fixed group-number, and then of counting the groups in the same way, the basis was laid for an extensive number system in which three developments were immediately possible:

[*i*] A rational and clearly understood number language could be built upon a limited number of words.

[*ii*] Reckoning could be greatly extended.

[*iii*] A written system could be devised with an even smaller number of symbols.

Just how and when these ideas evolved must always remain obscure but we know that by about, say, 2000 B.C., several number languages and systems of written symbols were in use in, for example, Egypt, Palestine and Babylon, and methods of reckoning were advanced enough to meet the needs of daily life in flourishing civilisations.

It is important to note here that the development of a system of reckoning beyond simple counting could occur quite independently of the existence of any system of written notation. Indeed, it is probable that the normal and natural course would be for the written notation to follow the development of practical reckoning methods.

The known history of events in Greece and Rome leads to this conclusion, and there is abundant evidence that counter-casting was widely practised in medieval and later times by people who had not yet learnt to write or read. As John Awdeley said in 1574, 'For as much as there be many persons that be unlearned, and can not wryte, yet nevertheless the craft or science of Awgrym and reckoning is needful.'[2] *See* illustration on p. 50.

Evidence suggesting the use of counters in early civilisations, possibly before the introduction of written notations, comes from the finding of small flat discs, usually of stone, at various levels during excavations in Palestine and Mesopotamia. Professor R. A. S. Macalister has mentioned[3] groups of pebbles found at Gezer which had been used, he suggested, 'either to serve as draughtsmen or perhaps to assist calculations like the pellets of an abacus'. He also found several groups of sheep astralagus (ankle) bones polished smooth on the two plane sides, which might well have been used as counters. It may be no mere coincidence that groups of similar bone counters have been found, more than once, on Saxon sites in England.[4] The same kind of bones could have served the same purpose two or three thousand years later.

Small stones and discs have often been found on other sites, for example Jericho[5] and Kish,[6] and a variety of possible uses for them have been suggested, any of which could be true. It has sometimes been said that they were gaming-counters, and this possibility is supported by the occurrence on some sites of tablets clearly marked for board-games.[7] There is no doubt, of course, that games in which 'men' were moved in spaces marked by lines on boards or stone tablets were played in very early times; but it is equally certain that the more practical side of life, even in those days, required a fair amount of calculation.

It may well be that these board-games bore the same relation to the more serious occupation of reckoning with counters that the rather similar Greek and Roman board-games did to the use of the abacus, and that the seventeenth-century game of 'shuffle-board' did to counter-casting.[8] As H. R. J. Murray has said in *A History of Board Games other than Chess*, 'A board, even if it shows the same pattern of lines or cells as a known game board, may have served other purposes before it was associated with a game. . . .'

Although we have been mainly concerned, here, with developments leading to the use of the abacus, we must remember that there have always been other, more direct, ways of solving problems involving computation, or of making it unnecessary. If the need for something more than counting and the memorising of simple number combinations arises, a practical method can sometimes be used. For example, the lengths of two pieces of wood can be added by laying them end to end, or their difference can be found by putting them side by side. Measuring ropes or tapes can be used in similar ways. Weights and money values can be compared, multiplied and divided by placing the actual weights or coins on a table, and other measures can often be treated similarly. It is possible to work out quite complicated tables of numbers, for example squares and square roots, with no more apparatus than a few pebbles or a measuring rod.[9]

The mathematics of Ancient Egypt was of this empirical nature.[10] The early Egyptians appear to have had no number theory that would enable them to calculate with written symbols, and there is no evidence that they knew the use of the abacus. The reference by Herodotus, in the fifth century B.C., to the use of pebbles by Egyptians has been taken as an indication that they knew how to use the abacus; but few of the discs or counters from excavations in Egypt are of a shape and size that would make them suitable and convenient for use in reckoning. Some are obvious gaming-counters, having been found in association with dice or gaming-boards; and there are wall illustrations that show such games in progress. Among the many wall illustrations showing various aspects of daily life there is no sign of an abacus.

Egyptian computation did not extend much beyond doubling and halving, but, as the Rhind Papyrus shows, quite advanced problems were solved by practical methods.

A clue to the way in which computations may have sometimes been managed lies in a curious dot diagram which appears on the back of a papyrus of about 1500 B.C., now in the British Museum. Another papyrus has a similar diagram.

It consists of ten rows, each of ten dots, with a line below the fifth row. An incomplete, separate, vertical row appears at one side. The diagram appears to be a casual one unconnected with the picture of the bull below it, or with the contents of the papyrus.

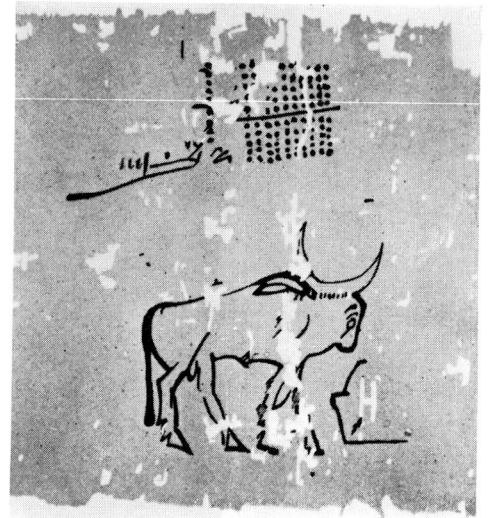
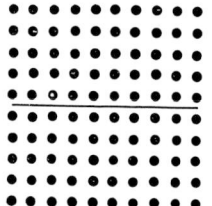

Papyrus 10184 (Salier 4), British Museum

It could, perhaps, be an 'aide-mémoire' for an Egyptian engaged in the tedious business of calculation. Addition and subtraction of numbers below a hundred could be effected by counting forward or backward along the rows; and the diagram could be used as a form of multiplication table. Similar diagrams are sometimes used today in primary schools.

Since no more than two examples of the Egyptian dot diagram are known it cannot be suggested that there was more than occasional use of such a device; but their appearance fits in well with the belief that Egyptian calculators of the time used a combination of mental arithmetic and practical methods.

Their notation, like most others, could not easily be used for the purpose of calculation, but served admirably for the recording of the numbers concerned. This is probably true of the cuneiform system and is certainly true in the case of Roman numerals. 'Arabic' notation, in fact, is the only system in which the written figures themselves can be used conveniently and efficiently in performing actual calculations, but whether or not it was invented, or it evolved, specifically as a method of calculation as well as a means of recording remains a mystery. It was generally known as 'Arabic' in Western Europe because knowledge of it spread through the teaching of Arab scholars, but the system undoubtedly originated in India. There is uncertainty about the date of the origin of the Hindu characters; and there is no evidence as to whether or not it was preceded in India by the use of the abacus method.[11]

We can be more certain about methods of calculation used in Ancient Greece because detailed records, inscribed on stone, have been preserved, and several specimens of the actual abaci used have been found. Amongst the earliest Greek records are those of Herodotus (485–425 B.C.) who not only mentioned the use of pebbles for calculation but gave numerous examples of mathematical problems. Some of these calculations of the fifth century B.C. are remarkably similar to exercises still appearing in school textbooks of the twentieth century A.D.!

An example quoted by Herodotus was to calculate the interest due after 1464 days on a principal of 766 talents, 1095 drachmas, 5 obols, at the rate of 1 drachma per day on 5 talents. (1 talent = 6000 drachmas; 1 drachma = 6 obols) Professor Mabel Lang of Bryn Mawr College, Pennsylvania, has shown[12] how the necessary computations for this and similar problems could have been carried out on an abacus of the Salamis type.

It is to be remembered that unhurried, methodical use of an abacus is by no means as difficult or complicated as it may appear to be if all the moves are described in detail in words, diagrams or written notations. Using an abacus is much easier and quicker than explaining how to do it. However, this example and others from inscriptions of the fifth century B.C. show that the abacus was already at quite an advanced stage of development at that time.

A complicated system of finger signs was used in Greece at one time, but it could only be used for the simplest of calculations. For more difficult ones the abacus was used.[13] At a later stage quite complicated calculations were worked laboriously with written symbols using a system based upon the letters of the alphabet. A number of such calculations were set out at length by Eytocious of Ascalon in his notes to Archimedes in the third century B.C. One of them was to multiply the number $3013\frac{3}{4}$ by itself; and the correct answer, $9082689\frac{1}{16}$, was reached. Addition and subtraction, however, were always done on the abacus at this time, and so were multiplication and division by low numbers; but, as a general rule, more difficult multiplication was done with written signs, and division by a combination of both methods.

Use of the abacus method passed in due course from Greece to Rome and thence to those parts of Europe influenced by Roman customs, including, of course, our own country. The British peoples living here before the Roman invasion had no written numerical notation of their own,[14] and there is little direct evidence of any method of calculation they may have known. Both in Gaul and in Britain Latin was used where Celtic people found it necessary to record names, on coins for example, and Roman appears to have been the only numerical notation they knew. A bronze calendar of the late first century B.C., found at Coligny in France, which is the oldest extensive example of writing in a Celtic language,[15] has Roman lettering and numerals.

However, there would have been comparatively few occasions when a written language was needed, whereas there must have been many when oral counting and reckoning came into everyday transactions. The use of currency bars during the second century B.C., and the Celtic coinage in gold, silver and bronze, show appreciation of scales of value. There was a certain amount of trade between Britain and the Continent, as well as locally, and a unit of value, the 'cumal' was in common use. (A 'cumal' was originally a female slave and was considered to be the equivalent in value to six heifers or three milch cows.)[17]

Small calculations could have been managed without difficulty but there must have been occasions when the numbers, or measures, were too large to be carried in the head, or dealt with by practical methods. The presence of numbers of flat discs of stone, pottery or clay on many Iron Age sites suggests that these people, too, may have had some kind of abacus method.

As the Roman way of life became established in Britain, Roman measures, money values and numerical notation came into general use and they are still with us. Such common measures as the mile and the pound are of Roman origin, and so, of course, is the abbreviation £ s. d. (libra, solidus, denarius).

Roman notation is still frequently used as an alternative to Arabic for some purposes; but the fact that we can interchange the two systems illustrates the important point that what is written is not the actual number, which is an abstract concept, but a visual representation of it. A parallel is found in music. Various notations can be used in recording music on paper; but the music itself exists quite independently of the way in which it is represented graphically. Notations are languages without which it would be difficult to communicate the ideas they represent.

Numerical notation has taken many forms in the past, including cuneiform, hieratic, Greek letters, Roman numerals and Hindu-Arabic figures. Changes from one to another have not necessarily meant changes in mathematical thinking; and a method of reckoning such as the use of pebbles on an abacus may have continued, with little change in its basic principles, throughout the long period during which a succession of systems of notation has been used, and, indeed, for a considerable time before any of them was devised.

The importance of distinguishing between mathematical ideas and the notations by which they may be expressed was emphasised by Cajori[17] when he said, 'Had the Greeks not possessed an abacus and a finger symbolism, by the aid of which computation could be carried out independently of the numerical notation then in vogue, their accomplishment in arithmetic and algebra would have been less than it actually was.'

Some of the words concerned with this part of mathematics are commonly used in ways that show that this distinction is not always appreciated. The word 'number' itself has come to have a confusing variety of meanings. As a noun it can mean 'a sum of abstract units' or 'a figure or symbol representing an arithmetical total' (*Shorter Oxford Dictionary*)' It is sometimes used, rather vaguely, for the study of numbers and their relationships, and it is often used in primary schools for any kind of introductory arithmetic, theoretical or practical or a combination of the two.

Written symbols, or characters, may, quite properly, be called either 'figures' or 'numerals'. The word 'figure' has other meanings, but 'numeral' refers only to a number symbol. 'Numerical notation' is, perhaps, the most explicit term. One may speak without ambiguity of Roman or Arabic figures, numerals, notation or numerical signs, but not of Roman or Arabic *numbers*.

A 'number' is an abstract concept which may be used quite independently of the language by which it is described or of the way in which it is represented in writing. For example, the number *ten* can be written as

It can be represented by a notch on a tally-stick or by a coloured rod, by a group of counters in one position on a board or by a single counter in another. Any of these written signs or concrete representations can be spoken of as 'ten' (or the corresponding word in another language), and can be used to express thoughts about the number; but 'the number ten' itself remains an intellectual concept.

It follows that to learn a system of notation or symbols does not imply an understanding and appreciation of the numbers concerned.

'Compute' and 'reckon' are almost synonymous, although

reckoning may require little more than simple counting, whereas computation implies some degree of calculation.

The word 'calculation' itself has come to have several shades of meaning, both in mathematics and in more general contexts, but its original association with pebbles on the abacus is now largely forgotten. It will be useful to consider the circumstances in which it originated.

II

CALCULUS AND ABACUS

'. . . the abacus on which they calculate'

EUSTATHIUS

Frequent references were made by contemporary writers in Ancient Greece and Rome to the use of pebbles, or other counters, for reckoning. Demosthenes, for example, in the fourth century B.C., referred to the need to use pebbling for calculations too difficult to be done in the head; whilst a character in a comedy by Alexis (4th–3rd century B.C.) called for an abacus and pebbles to do his accounts. Both Diogenes (412–322 B.C.) and Polybius (*c.* 204–*c.* 120 B.C.) spoke of men who sometimes stood for more, sometimes for less, like the pebbles on the abacus.

The Latin phrase 'ponere calculos', literally 'to place pebbles', meant 'to reckon'; and 'calculum subducere' was used by Cicero (106–43 B.C.) in the same sense. He also used the phrase 'ad calculos vocare aliquid', meaning 'subject to strict reckoning'.

The literary references are suggestive of a general use of the abacus method of reckoning in those days, but a more immediate and obvious indication lies in the modern English word 'calculate' itself (and the French 'calculer'). It is clearly derived from the Latin 'calculus', a pebble, the diminutive of 'calx', a stone. 'Calx' was used especially for limestone; 'chalk' is derived from it, and so is the name of the metal 'calcium', one of the constituents of limestone.

The modern use of the word 'calculus' for a branch of mathematics in which the methods of calculation are largely theoretical shows a curious transmutation from the original, very practical, meaning; but even more strange is the chequered history of the word 'abacus'. This word is used today, rather loosely, as a general term for any calculating device in which beads are moved on wires or in grooves, as well as for the lined board or table on which Greeks and Romans manipulated their pebbles. It is also used for the bead-frame computers which are still in general use (although they are not, in fact, very ancient) in China, Japan, Russia, Turkey, and some other Eastern countries.

Extension of the word to these devices which seem rather complicated and a little mysterious, to Western eyes, is unfortunate, for it hides the essential simplicity of the original meaning and of the method of calculation with which it was associated.

The Roman abacus, like its predecessor in Greece, was no more than a convenient flat surface upon which pebbles could be placed. This was the original meaning of the word. The Latin 'abacus', derived from the earlier Greek word 'abax' (ἄβαξ), meant, simply, 'a flat surface'. It was used for any kind of table or tablet, dining-table, side-board, writing-tablet, gaming-table, a tray for kneading dough, as well as for a reckoning board or table. A flat panel decorating a wall was called an abacus, and so was (and still is) a flat stone slab used as an architectural feature such as the stone square at the top of a pillar. A now obsolete English word 'aback', used by Ben Johnson, meant an ornamental square panel.

The diminutive 'abaculus' (ἀβακίσκος) was used for an ornamental tile of marble or glass.

It is frequently stated that 'abax' is derived from an ancient word of Semitic origin, the Hebrew 'abaq', dust, a less common synonym of the ordinary Hebrew word 'apar'; but the relation between the Hebrew and Greek words is not at all certain.[1] According to one authority it is 'nicht zutreffend' (not proven),[2] and there is little evidence to support a common idea that a table strewn with dust, or sand, was at one time widely used for reckoning.

Certain references in classical literature seem to suggest that sanded tables were used for reckoning; for example, the Roman satirist Persius (A.D. 34–62) spoke of 'the sort of person ... who knows how to laugh slyly at numbers on the abacus or goals in the divided dust'. (The goals, 'metae', were normally the pillars marking the two ends of a chariot-race course.) Sanded tables certainly seem to have been used for the drawing of *geometrical* figures,[3] and if the sand was damp or nicely smoothed such an abacus would have been suitable and convenient for this purpose. But it is not so easy to imagine counters being moved easily from place to place on a sandy surface, and grooves would only add to the difficulty of moving them. A hard surface of dried mud seems possible, but any loose sand or dust would be a nuisance. The flat shape, especially the underside, of the known counters and the later use of metal discs show that they were intended to slide from place to place (hence the French name 'jeton') and should not be picked up. In any case, there would be no reason for strewing sand on a table surface when all that was needed was two or three chalk lines or scratches.

Further, it is to be remembered that the method of calculating with pebbles is ancient enough to have been invented in some Eastern country at a time when it was still the normal custom to sit on the ground rather than at a table. A carefully smoothed patch of ground or rock could have served as a reckoning 'board', and it would then have been natural enough to speak of reckoning on the 'abaq'. The custom of using sand or dust *may* have been transferred to tables when they came into use, but it seems more probable that it was the name only that was transferred from the one flat surface to the other.

The thesis that the normal method of calculation in Ancient Greece and Rome was by moving counters on a smooth board or table suitably marked with lines or symbols to show the 'places', is supported by ample evidence of a more concrete character than literary references. Although no wooden boards have survived from these times, several stone or marble Greek abaci have been found (*see* appendix) and suitable counters, especially Roman calculi, are found on many sites; and the existence of two or three illustrations of calculators actually using pebbles places the matter beyond doubt (*see* pp. 26 and

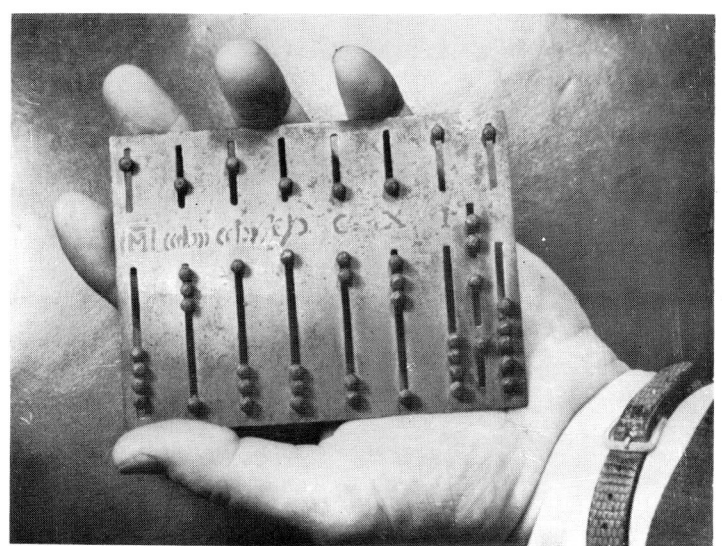

Roman 'abacus' in the British Museum

27). There is also strong evidence to show that medieval reckoning, about which a great deal is known, is based directly upon, and evolved from, the Roman practice.

Before discussing the archaeological evidence, however, it is necessary to refer to a small number of remarkable 'bead-frame computers' of Roman origin to which the name 'abacus' has also been given. These ingenious 'pocket calculators' have tended to divert attention from the normal Roman method of calculation with loose counters, but since only a small number of these devices have been found, they cannot have been in general use.

One of them is in the British Museum, another is in the museum of the Bibliothèque Nationale in Paris and a third is in the Museo Nazionale Romano. One other has been described but is now lost, and it is possible that there was a fifth. All are similar in form and size. They are small enough to be held in one hand while the beads are moved in the slots with the other.

Little or nothing is known about the origin of any of these

C

instruments and there is, therefore, no possibility of attributing a definite date to them; but there can be little doubt they are indeed Roman. The markings of the columns suggest an early Roman date; and the fractions, halves, thirds, quarters and twelfths, correspond to Roman money values. If these were medieval devices, as has been sometimes suggested, there would surely be provision for shillings (or sols).

The similarity of the Roman bead-frames to those used in some Eastern countries today (*see* p. 100) suggests that the use of such instruments spread in some way from Rome to China and thence to Japan and Russia. It would be difficult, however, to establish any direct connection, for little is known of the early history of the Chinese 'suanpan'. It was mentioned once or twice in earlier histories of Chinese mathematics, but rules for its use did not appear until the thirteenth century. The Japanese 'soroban' did not come into popular use until the seventeenth century.[4]

The part played by the bead-frame type of abacus in Roman life can only have been a minor one. They are ingenious, well-made instruments, but nevertheless a handful of suitable counters on a table could perform the same functions; and, because the loose counters have to be actually placed on the table and removed from it, the various moves of the operation are more easily followed and understood. Most significant, of course, is the very small number of bead-frames that have been found, whereas calculi are found on many Roman sites.

These are flat or plano-convex (bun-shaped) discs, usually $\frac{1}{2}$ to $\frac{3}{4}$ inches in diameter and may be of bone or ivory, but are often of glass—made, presumably, by dropping molten glass on to a flat surface. They are often white or black, but may be green, brown or blue. Some are decorated with three or four coloured spots. Rounded sherds of pottery are sometimes found.

In some museums calculi are shown described as 'gaming counters'. Board games were popular among the Romans and counters would certainly be needed for them. A set of thirty 'bicolores', fifteen of each colour, were found at the Roman villa at Lullingstone together with traces of the brass corners

Roman calculi

of a gaming-board, presumably of wood, nineteen inches square. The game may have been similar to backgammon.

The term 'ludus calculorum' was used for a Roman board-game played with calculi as 'men'. The Greek game 'pettia' was similarly derived from the use of pessoi (ψηφοί) on the abacus.[5]

Mere pastimes, however, would not account for the presence of calculi on so many Roman sites, or of other kinds of counters on so many earlier sites, whereas a certain amount of reckoning would occur wherever there was any kind of trade or industry involving the use of money and measures.

The Latin word 'calculus' survived for many centuries after metal 'jetons' had replaced 'pebbles' as reckoning counters. A jeton of Louis XII (1498–1515) has the inscription CALCVLI AD NVMERANDVM, and Silicius used the phrase 'calculus suppatorius' in 1526. A Belgian jeton of 1609 is inscribed CALCVL . CIVIT . BRVXELL; and the reverse of a jeton of

Jeton of Philip II used in the Spanish Netherlands

Philip II of Spain has the interesting legend CALCVLVS CAMERE RATIONV : INSVLEN and the date 1579.

'Camere' in this context meant a room, and 'camere rationu' a counting-house, or reckoning-room. 'Insulen' referred to the town of Lille. This is, therefore, a counter of the counting house of Lille.

A jeton of Louis XIV (1661–1715) bears the phrase CAL-CVLA . NE . DECIPIARIS which may mean 'Make your reckoning in order not to be deceived'.

Although Roman calculi are so frequently found, no example has yet appeared of a Roman table abacus. Several Greek abaci are known but the absence of Roman abaci need cause no surprise or misgivings. If wooden boards or tables were used they could not have survived, and any distinguishing marks on a stone or marble abacus may have disappeared. There is, in fact, an exactly similar dearth of reckoning-boards or tables although they were undoubtedly in common use little more than two hundred years ago. Any such tables that have survived have, with only a very few exceptions, lost their distinguishing marks through wear and tear of the surfaces during two or three hundred years of use for general purposes.

An ancient practice of using, occasionally, large, heavy marble or stone abaci, possibly for public use much as weighbridges are still used, may account for the discovery of twelve examples of such tables, or fragments large enough to recognise. In addition, some 'casually converted roof tiles' inscribed with numerical symbols, that may have been used for the same purpose, are also known.[6]

The most complete, and best known, of the Greek abaci is that found on the island of Salamis. Another has similar sets of parallel lines and five others show lines of some kind. Nine of them show a system of notation that could have been used for numbers or for money values. Little or nothing is known of associated datable material, but the inscriptions, together with the general circumstances of their discovery, indicate that these abaci were used some two thousand years ago.

The Salamis abacus is of white marble, 1·49 metres long, 0·75 metres wide and varying in thickness from 0·045 to 0·075 metre. It has two sets of parallel lines, one of six and the other of eleven lines. Numerical symbols appear on three sides, ranging in one case from 1 to 1000 and in another from one-eighth of an obol to 6000 drachmae. (Drachmae and obols were units of weight as well as money in ancient Greece, the obol being one-sixth of a drachma.)

The abacus is of a convenient size for use with pebbles which could have been placed on and between the lines as well as opposite the symbols. A detailed description of ways in which such an abacus could have been used for quite complicated calculations, with actual examples from Greek records, has been given by Professor Mabel Lang (*see* p. 11).

She suggests that the amounts concerned were shown by pebbles placed beside the corresponding symbols at the edges

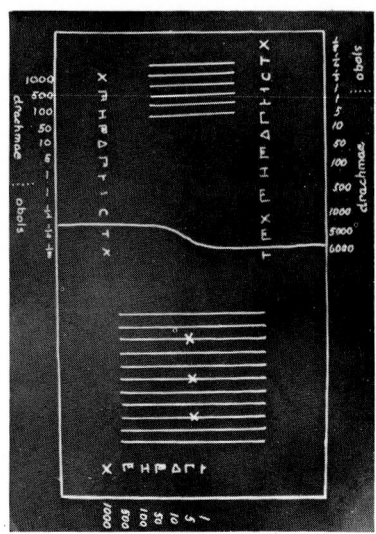

The Salamis Abacus and (right) diagram of the same

of the abacus and that the lines in the centre were the working area.

Detailed descriptions of the moves made on an abacus tend to make the processes appear much more complicated than they really are. The actual working time is much shorter than the time needed to explain it. Most of the calculations needed in the course of normal daily life would have been fairly simple ones, but complicated calculations could be managed quite well.

A number such as 16,789 could have been shown on the Salamis abacus as in either figure on the opposite page.

It would have been a fairly simple matter to add to the number, subtract from it, multiply it or divide it. The examples given by Herodotus show that the calculators of nearly two thousand years ago dealt successfully with problems involving computations of this order, although, no doubt, they took their time about it.

It is a curious fact that although the use of this method of calculation by the Greeks is well attested, especially by Herodotus, and supported by the discovery of several examples

consider together the various literary references, the deriva-
tions of some of the words used, the two or three illustrations
that have survived and the archaeological evidence, we must
conclude that there was general use of the simple tabular abacus
in classical times. Interesting confirmation of this is given by a
marble sculpture of the first century B.C., now in the Museo
Capitolino in Rome. A dying man is shown dictating his will,
while his wife makes notes beside him and a slave uses an
abacus to check that the bequests do not outrun the estate.

III

ROMAN NUMERALS TO ARABIC FIGURES

'I have often admired . . . the secret magic of numbers.'
SIR THOMAS BROWNE (1605–1682), *Religio Medici*

Arabic figures were first known to scholars of Western Europe early in the eleventh century, but their use spread very slowly and Roman numerals continued in general use in the keeping of accounts for several hundred years. Reckoning was dealt with by counters of metal (jettons) in place of the earlier 'pebbles' or calculi, and this familiar, traditional method coninued for a hundred years or more after the new Arabic system had been generally adopted, in their record books, by merchants and accountants.

There has been a good deal of speculation about the origin of Roman numerals and, occasionally, some wild guesses with little or no evidence to support them; but there is no doubt that, in the main, the Roman system evolved from the Greek. The earliest Greek custom was to use upright strokes for the smaller numbers; but two other systems were used in later times in different parts of the Greek Empire. Letters of the Greek alphabet, with a few additional characters, were sometimes used, but the most general custom was the one shown on the Salamis abacus, of using the initial letters of the words for ten, hundred, thousand and ten thousand, together with a single stroke for one:

Δ	initial letter of	ΔEKA	(deka)	ten
H	initial letter of	HEKATON	(hekaton)	hundred
X	initial letter of	XIΛIOI	(chil'ioi)	thousand
M	initial letter of	MYPIOI	(myr'ioi)	ten thousand

Roman calculàtor

Nationale, Paris. It shows a 'calculator' in exactly the same situation as the receiver of tribute on the Darius Vase; and, incidentally, as the 'Rechen-meister' on sixteenth-century 'rechen pfennigs' (*see* pp. 81 and 82). The writing tablet in his left hand shows the signs for thousands, hundreds, tens and units ($\phi, \oplus, \times, \uparrow$) while he is placing the calculi on the table abacus with his right hand.

This 'pierre gravée' is shown, somewhat inaccurately, in Darenberg and Saglio's *Dictionnaire des Antiquités grecques et romaines* of 1877. In the same book, under the entry 'ludi magister', there appears an amusing contemporary Roman caricature (of unstated origin) of a school in which there are six small boys, a dunce, a naughty boy and a pupil teacher; and a master with the head of an ass. Each of the pupils has before him what appears to be a double abacus, the upper half for

Caricature of Roman school

writing and the lower for calculation . . . 'l'abaque, table de pierre, de bois ou de metal, qui servait avec les cailloux ou calculi aux exercise de calcul . . .'

Primary education of boys in both Greece and Rome included, besides reading and writing, a certain amount of arithmetic, using the abacus. Plato suggested 'as much as is necessary for the purposes of war and household management and the work of government . . .'. The pupils, according to Carcopino,[7] 'spent hours counting the units, one, two, on the fingers of the right hand, and three, four, on the fingers of their left, after which they set about calculating tens, hundreds and thousands by pushing little counters, or calculi, along the corresponding lines of their abacus'.

It may be because arithmetical calculation was mainly utilitarian and a comparatively minor factor in the everyday lives of ordinary citizens that we have few direct references to the methods used, or clear descriptions of them. However, if we

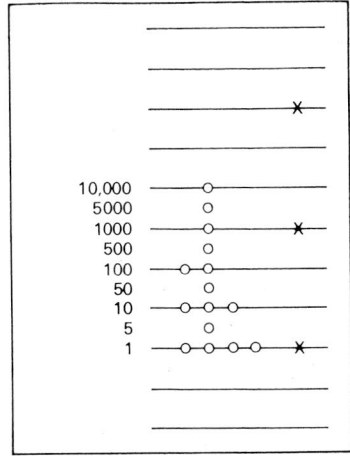
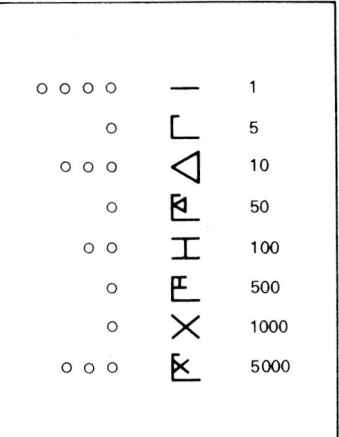

of the kind of abacus actually used, there appears to be no certainty about the kind of counters used. The name 'pessoi' (ψῆφοι) was used for pebbles of any kind as well as for reckoning counters; and the word ψηφίζω meant 'to count or reckon' in exactly the same way that the Romans later used the phrase 'ponere calculum', and that we, ourselves, now use the word 'calculate'. But, whereas Roman calculi are well known and easily recognised, the kind of 'pebble' used by the Greeks is uncertain. Various discs and other pieces, including those called 'stoppers' by archaeologists, are found on many Greek sites, but there appear to be none that can be clearly identified as reckoning counters.

However, there is no reason to suppose that only one kind of counter was used; and ordinary pebbles could have served.

The use of the table abacus in both Greece and Rome in conjunction with written recording of the amounts concerned is clearly shown in two illustrations that have survived, one Greek and one Roman.

A detail of the 'Darius Vase' (now in the Museo Nazionale di Napoli) shows the treasurer of the Persian king, Darius I, who conquered Greece about 500 B.C., receiving tribute. He is

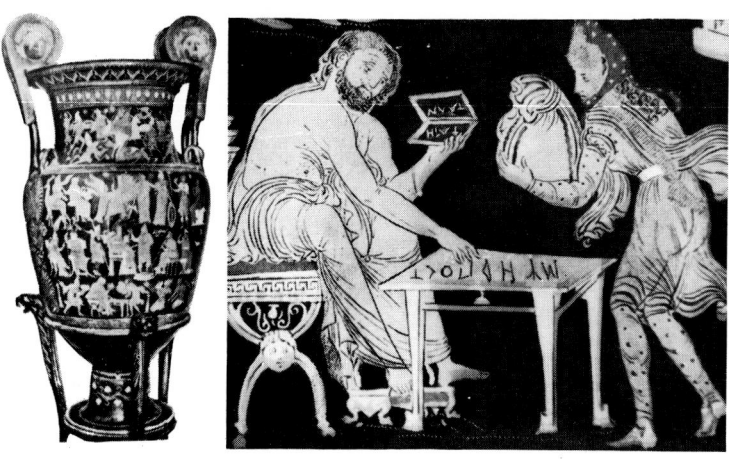

The Darius Vase—an Apulian vase of about 300 B.C.
Height 1.3 m. Detail (right), from Griesche Vasenmalerei
(*Furtwangler and Reichold, 1909*)

seated at a small table manipulating small white pebbles with
his right hand while holding a writing tablet in his left hand.
The tablet shows symbols which may indicate 'talanta hekaton',
i.e. 100 talents, but the symbols on the table range from
10,000 drachmas down to quarter-obol. The pebbles on the
table represent the sum 1231 drachmas 4 obols.

	○	○ ○	○ ○	○	○ ○ ○ ○			
M	**Ψ**	**H**	**Δ**	**Γ**	**O**	**ᒣ**	**T**
10,000	1000	100	10	1 (or 5?)	1	$\frac{1}{2}$	$\frac{1}{4}$
	Drachmas					Obols	

The Roman illustration appears in the form of an engraved
seal now in the Cabinet des Médailles of the Bibliothèque

To these were added:

Γ an old form of Π from **ΠΕΝΤΕ** (pen'te)   five

and the combinations:

Γ̄ For fifty   Γ̄Η For five hundred

The sign for one became ⊣ if used for an obol, and ⌐ if used for a drachma.

The Roman I needs no explanation; and the X, for ten, has none. It may have been merely an easy mark to write or engrave. C, for one hundred, may have come from the initial letter of 'centum', but it is more probable that it came from the older form ⌐, with L for fifty as the lower half of this sign. Another hundred sign, the Etrurian ⊕, later written as ⊿, which was still in use in English manuscripts of the sixteenth century, for example, the Windsor Castle accounts for 1540, shown on page 44. A possible connection between C and a quickly written ⊄ is obvious.

The origin of the thousand sign, M, is even more dubious, and it may be mere coincidence that this is the initial letter of the Roman word 'mille'. An older Roman symbol for a thousand was taken from the Greek letter phi, ⊄. It was written or engraved in various ways, sometimes in the form CD which has been used as recently as 1845 on a wall tablet on a staircase in the Ashmolean Museum, Oxford, where the date is engraved as CD D CCC XLV. It is also to be seen in the printed date CD DCCXLI (1741) of a *Codex Antiquissimus* in the Royal Malta Library and in the date on a Dutch commemorative medal of 1592 (*see* p. 32). It appears in the cursive form ⊄ on the Roman beadframe in the British Museum, and this is extended to ⟨⟨⟨⟩⟩⟩ for ten thousand and ⟨⟨⟨⟨⟩⟩⟩⟩ for a hundred thousand.

It is quite possible that the freely written ⊄ may have developed into the written ⊓ and then to the capital letter M. D, for five hundred, may have come from the sign D; but it could be claimed that this is the initial of a Latin word, demi.

Possible explanations have already been given for the origin of the signs for five hundred (D) and fifty (L) but there is no

Dutch commemorative
medal of 1592

such obvious explanation of the V sign for five. Two or three possible origins have been suggested, but without definite evidence it is impossible to choose between them.

A more interesting question is why they (V, L and D) were used at all. It is important, here, to remember that the use of the abacus began long before the written systems were devised. This is certainly true of the Roman and Greek systems and is probably true of all other written systems, including, of course, the Hindu system on which the so-called 'Arabic' is based. It was for the more efficient use of the abacus method that intermediate values were needed.

This can only be fully appreciated by actual practice with counters. If one tries to operate with up to nine counters in each 'place' (whether on the lines or between them) there is soon some uncertainty and confusion in quickly recognising the numbers indicated. For example, to show the number 9968 one would have on the abacus (using a possible Roman arrangement with units, tens, etc. between the lines):

M	C	X	I
○ ○ ○ ○ ○ ○ ○ ○	○ ○ ○ ○ ○ ○ ○ ○ ○	○ ○ ○ ○ ○ ○	○ ○ ○ ○ ○ ○ ○ ○

It is difficult to see at once what the numbers are (unless the counters are arranged very systematically, which would not be easy); but if intermediate places are used for five, fifty and five hundred the numbers can be read quickly. There are never more than four counters to take in at a glance.

M	C	X	I
○	○	○	○
○	○		○
○	○		○
○	○		

The probability is that intermediate places were used at an early stage, and that the D, L and V were added later to the original M, C, X, I symbols.

It may be no more than a coincidence that I is written, or engraved, as a single stroke, V, X and L with two strokes, the old [(hundred) with three strokes, and M with four strokes.

All the common numbers or values could be written quickly and easily, on a waxed wooden tablet using capital letters in Roman times, and nearly fifteen hundred years later using the 'italic' script on parchment or paper. It is only when large numbers are involved that the Roman system becomes clumsy, and there could have been little need for large numbers for everyday Roman life. The normal custom was undoubtedly to use a writing tablet for setting down the figures to be dealt with and then the answer when it had been obtained, the actual computation being effected by the manipulation of calculi. This, incidentally, is once again normal business practice, with a modern calculating machine in place of the abacus.

For the keeping of records of all kinds, the writing of dates, numbering the items in a series (e.g. the pages in a book) and book-keeping, the Roman system proved to be so convenient and suitable that it was used exclusively in Europe for well over fifteen hundred years, and, in fact, it is still quite widely used for some purposes. To us it may seem clumsy, but there was no other system to challenge it until the coming of the Arabic notation.

The Hindu-Arabic system, with a different symbol for each of the nine integers plus a sign for zero, has the great advantage that it is more neatly written, especially with the larger numbers, and can be extended indefinitely to deal with very large numbers in a comparatively small space. It also has the advantage that 'place value' is used for the written numbers and so they correspond to the numbers of counters in each place if the abacus is being used. (In the first recorded use of Arabic notation in Europe, Gebert, who became Pope Sylvester II in the year A.D. 999, used an abacus arrangement with Arabic numbers written on counters.)

It may seem surprising to us who are familiar with the Arabic system that it was not readily adopted as soon as it was known, but there were good reasons why the change over from Roman to Arabic took place very slowly. In the first place, there were very few people who could read or write until the seventeenth and eighteenth centuries. Pen-reckoning and even the figures themselves were treated with suspicion; indeed, there are still those who say, with Thomas Carlyle, '. . . you might prove anything by figures'.

In the year 1299 an edict was issued in Florence forbidding bankers to use the new figures, and in 1348 the University of Padua directed that a list of books for sale should have the prices marked 'non per cifras, sed per literas clara' (not by figures, but by clear letters, i.e. in Roman numerals).[1]

There was also the mysterious symbol 'o' that appeared among the Arabic figures, affecting their values but having no value in itself. Although the use of the zero sign was sometimes understood in medieval times, for example by William Langland who wrote in 1399 '. . . as sifre doth in awgrym, That noteth a place, and no thing availith'[2], it was, for most people, a confusing conception that did not arise in their use of the counter-board or in the writing of Roman figures. There was not any difficulty over the lack of a zero, as has sometimes been suggested; on the contrary, an empty space on the board made a calculation so much easier and the answer, in Roman figures, so much shorter to write.

The difficulty and confusion of thought associated with the notion of zero added considerably to the delay in the general use of Arabic figures for calculation in the seventeenth and eighteenth centuries, and even today we use the word 'cipher' (derived from the old Arabic word for zero, 'sifr') in a somewhat capricious manner. It is still sometimes used in the original sense of zero, but it can also mean *any* Arabic figure! As a verb, 'to cipher', it may mean to calculate, but such is the mystery surrounding the word that 'to cipher' can also mean to use a secret code!

There may be an unintentional commentary on the situation in Bailey's *English Dictionary* of 1725: 'Ciphers are certain odd Marks and Characters in which Letters are written, that they may not be understood . . .'.

But apart from any difficulties in understanding the strange Arabic notation, the main reason for the continued use of the Roman system for some hundreds of years after the new figures had been first introduced into Western Europe, must have been that people were well satisfied with the traditional system. With counters to perform the actual computations, Roman notation was quite adequate. In one important sense, in fact, Roman notation was, in the eyes of the counter-casters, superior to Arabic, for every Roman figure corresponded to a counter, and vice versa (unless such abbreviations as IV, IX and XC were used).

Thus, the number 1786 would be shown on the abacus as

and written as MDCCLXXXVI. On the Arabic system the relation between counters and figures was not so easily seen.

The Arabic system of notation probably originated in India, but may not be as ancient as is sometimes thought. There are

D

records of some of the signs being used in the third century B.C., the first complete system of nine signs appearing in the first century A.D. The zero sign, however, did not appear until much later and the earliest undoubted occurrence of a zero was in A.D. 876 at Gwalior.[3]

It would not have been easy to use these Hindu signs for calculation without the zero sign, which, of course, implies an understanding of place value; but they were an improvement on Roman and all previous systems even for purposes of recording. They might well have been used at first in association with some form of manual calculation such as the abacus method.

The system was attributed to the Arabs by the scholars of Europe because it came via Baghdad with the Moorish invasion. When southern Europe was overrun by 'Saracens' during the seventh and eighth centuries A.D. they brought with them their learning, including the Hindu number notation which they had adopted. Later they founded schools and universities where it was possible for European scholars to acquire this new knowledge.

It has been claimed that the mathematician Boetius (c. A.D. 475–524) first showed the Arabic system to the Western world, just as Pythagoras is said to have introduced the abacus into Greece nearly a thousand years before. Boetius and Pythagoras are shown using their respective systems in an illustration to Reisch's *Margarita Philosophica*, published in 1503.

Boetius' part, however, is very doubtful and it was not until the tenth century that Arabic figures first appeared in European manuscripts. Their use grew steadily, but slowly, during the next two or three hundred years, not so much for the purpose of calculation, but mainly for recording. Arabic figures were clearly more suitable for such purposes as numbering the pages of a book, especially as the numbers grew larger; but calculation with them was an entirely different matter.

Some European documents of the thirteenth century had Arabic figures throughout. They appeared in a thirteenth-century manuscript on the new 'craft' of *Algorismus*, preserved in the British Museum, and were used in a German missal of

REISCH'S *MARGARITA PHILOSOPHICA*: 1503

about 1385, and again in Caxton's *Mirror of the World*, printed in 1480. The first coin to bear the date in Arabic was a silver gulden-groschen of Tyrol issued in 1486; but the first English coin to show a date, a shilling of Edward VI, gives it in Roman, MDXLVIII (1548). Later in this reign dates were shown in Arabic.

During the early years of the use of Arabic figures they were not infrequently mixed, or combined, with Roman; for example a French jeton of 1494 shows the date as MCCCC94. In a letter

DECIMAL FRACTIONS.

A decimal fraction derives its name from the latin *decem*, *ten*, which denotes the nature of its numbers, reprefenting the parts of any integeral quantity divided into a decuple, or tenfold proportion.

NUMERATION,

Teacheth to read or write any number propofed either by words or characters, according to the following

TABLE.

6 5 4 3 2 1·2 3 4 5 6

C.M. X.M. C. X. M. C. Units C. parts X.M. parts M. parts X.M. parts C.M. parts

C.M.

Roman numerals used to explain the place values of Arabic (From Leybourne's Cursus Mathematicus, *1690)*

to Lord Burghley written in 1589, the writer used '25th' in one place and 'xxvth' in another. Even as late as the nineteenth century people still occasionally combined the two systems.

Trust in the old system was shown in 1690 in Leybourne's monumental *Cursus Mathematicus* by the use of Roman numerals to clarify the still not entirely familiar Arabic system of notation.

It is interesting to see that the decimal fractions are shown as a logical extension of the Roman system.

A hundred years later W. Taylor, in his *Arithmetician's Guide*, 1793, still found Roman letters useful, at the same time showing the possibility of extending the Arabic system far beyond the range of the Roman.

The gradual change from Roman numerals to Arabic, without, at first, any departure from the traditional counter-casting, is well shown in the account books of the time. In the Rent Roll

TABLE II.

Periods	Quadrillons.	Trillions.	Billions.	Millions.	Units.
Half per	th. un.	th. un.	th. un.	th. un.	c.x.t. c.x.u.
Figures	345,432.	615,423.	689,345.	214,032.	324,516.

Arithmetician's Guide *by W. Taylor*, 1793

of St. Andrew's Chapter, Scotland, Arabic figures are used for the date 1490; but Roman characters, written in italic, were in general use until at least the end of the sixteenth century.

The Mayor's Audits for the City of Bristol show accounts entered in Roman only until late in the sixteenth century when Arabic began to appear occasionally. In 1635 the last two pages were completely Arabic, but there was a reversion to Roman the next year. From 1640, however, only Arabic was used.

It seems that here there was something of a tug-of-war between a conservative-minded clerk sticking grimly to the old method and a bright young man anxious to introduce the new fashion in figures. In the end, of course, the old order had to go.

Other documents in the care of the Bristol City Archivist confirm that the change from Roman figures to Arabic took place, in Bristol at any rate, between 1635 and 1640. Evidence that the new figures were known beyond the accountant's office is seen in a workman's account, dated 1639, written on a scrap of paper with the amounts in Arabic figures (but wrongly added!). Another, dated 1640, showed, on the back, an addition sum set out almost in the modern manner:

li	s	d
02	01	6
01	02	4
03	03	10

('li', of course, is short for libra, pound. It later became written as £.)

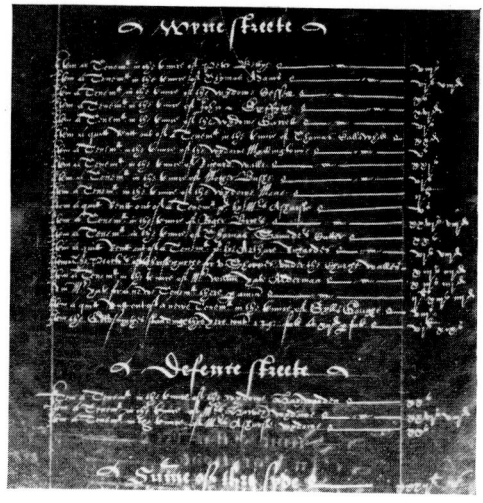

1599. *All entries and total in Roman.*
N.B. *Addition presumably by counter-board*

The entries in the column on the right are:

li	s	d	li	s	d	li	s	d
8			6			2	13	4
6	8		2					6
10				5		6	19	
	16		26	8				
6			20					
	20		6	8		20		
5			13	4		26	8	
21			13	4		20		

Sume of this syde—xxxij li v s (£32 5s) (Other entries included)

1640. *All entries and total in Arabic.*
N.B. *There is no indication here of how the addition was made*

A variety of documents published by the Bristol Record Society cover this period and show similar changes in the notations used. They include records of the Society of Merchant Venturers in the seventeenth century, and the Company of Soapmakers for 1562–1642, Apprenticeship Books for 1600–1689, Burgess Oaths for 1600–1698, Wills, Inventories and Shipping Surveys. Arabic figures were first used, about the middle of the sixteenth century, in writing dates, especially the year; and they were used, rather irregularly and sometimes mixed with Roman, in recording amounts of money from the beginning of the seventeenth century. All entries were made in Arabic only from dates, varying from one set of documents to another, between 1635 and 1650.

The last Roman entries in any of these documents occurred in 1661.

Customs and methods of book-keeping varied from place to place, of course, and the careful writing of accounts by hand was very much a personal matter. The change from the Roman system to the Arabic, therefore, was a slow and uneven one, but it appears that after the middle of the seventeenth century there were few keepers of accounts who had not recognised the advantages of the new system and changed to it.

At first, however, the change was only in the adoption of a neater and clearer system of notation in the making of records while the traditional method of reckoning with counters on the counting-board or Exchequer Table continued. The use of Arabic figures for calculation was another matter, much more difficult for businessmen and their clerks to grasp, and there must have been much discussion and argument during the next hundred years or so before the old method of counter-casting was finally abandoned.

COUNTER-CASTING TO PEN-RECKONING

'I shall reken it syxe times by aulgorisme or you can cast it ones by counters.'

J. PALSGRAVE, 1530

Some of the most interesting documents showing how accounts were kept in the sixteenth and seventeenth centuries are the Exchequer Records now in the Public Record Office. They show similar changes from Roman to Arabic figures to those described in the last chapter, but they also show in an unexpected and intriguing way that the Exchequer Table was in use even after the change from Roman to Arabic figures.

Arabic figures were first used towards the end of the sixteenth century and Roman numerals had entirely disappeared by the middle of the seventeenth. The Windsor Castle Accounts for 1540, for example, used Roman only (*see* p. 44), Arabic first appeared in 1597 and some pages for that year were in Arabic only, but it was about fifty years before Roman numerals had finally disappeared from the Accounts.

An interesting use of Arabic figures for numbering the items is shown in the second illustration (*see* p. 45) but, otherwise, both the pages shown have all amounts entered in Roman numerals. Both show correct totals but the lack of any orderly arrangement of pounds, shillings and pence indicates that they were not obtained by simple addition of the kind we are familiar with today.

In the third example shown (*see* p. 46) there is even an Arabic total at the bottom of a column of Roman entries:

This does not, however, mean any kind of confusion, for there is ample evidence that the totals were arrived at by the use of counters. A total was read off from the final position of the counters as they lay on the Exchequer Table, and recorded in whichever notation the recorder chose to use. 'Dot-diagrams'

Windsor Castle Accounts, 1540. Temp. Henry VIII. Roman figures throughout. (Public Record Office)

li	s	d	£	s	d
x			10		
iiij	xj	iij	4	11	3
iiij	xj	iij	4	11	3
iiij	xj	iij	4	11	3
	ℓj	iij		101	3
	ℓ			100	
xxxiij	xv	0	33	15	0

appear on several pages of the Exchequer books of the period and can be seen in the last two illustrations. It was suggested by Sir R. H. I. Palgrave in his *Dictionary of Political Economy*,

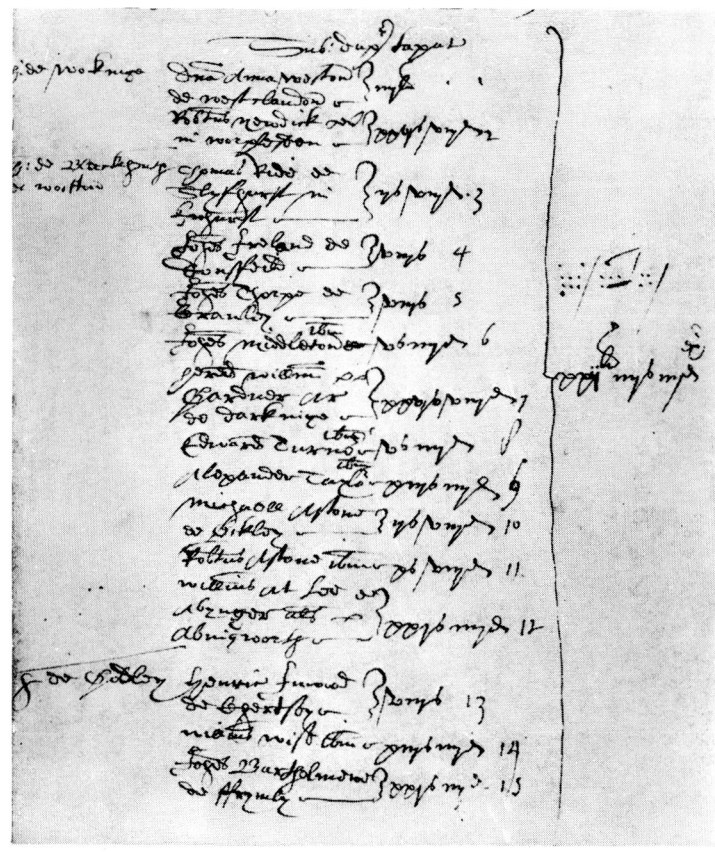

Exchequer Augmentation Office, Miscellaneous Books 1603. Temp.
Elizabeth. (Public Record Office)

1894, that 'In Tudor times pen and ink dots took the place
of counters . . .'; but since the dot-diagrams show the totals
only, and any attempt to compute with such dots would
lead to confusion, there can be no doubt that the diagrams are
clear evidence of the use of the Exchequer Table.

Interpretation of the dot-diagrams is not difficult and can be
deduced by examination of the three examples given here, or
understood by reference to the following diagram of the
positions of the counters on the Exchequer Table in what was

Exchequer Augmentation Office. Misc. Books. 1600. Temp. Elizabeth.
Amounts written in Roman with the total in Arabic.
(Public Record Office)

li	s	d	£	s	d
xlviij	iij	iiij	48	3	4
xvij	vij	iii	17	7	3
xxi	xix	v	21	19	5
xxx	viij	viij	30	8	8
xiij	viij	j	13	8	1
xxxij	ij	j	32	2	1
xxiv	xv		24	15	0
xxxix	vij	vj	39	7	6
			227	11	4

described by Robert Recorde in the sixteenth century as the
'Auditor's Use':

£20			£			s.			d.		
		100	10		5	10		5	6		
20	20	20	1	1	1	1	1	1	1	1	1
20			1			1			1	1	

In the example shown below, for instance, the total is shown in Roman (cursive) notation as lxx li. xiij s. viij d., i.e., £70 13s. 8d., although the dot diagram, in error, shows 9d.

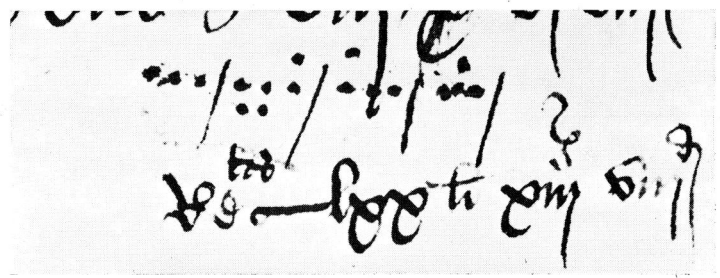

Detail from Treasury Account, T.R. Books. 1540. Temp. Henry VIII

The dot-diagram shows this amount as indicated by the counters lying on the table at the end of the operation:

			£			s.			d.		
					5	5			6		
20	20	20	1	1	1	1	1	1	1	1	
			1	1							

£70   13s.   8d.

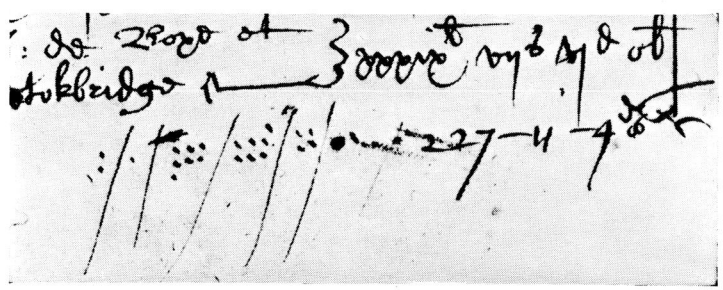

The dot-diagram in the detail from the document of 1600 seen above can be interpreted rather more easily because the total in this case is written in Arabic figures.

The first two dots clearly represent two hundreds (of pounds), and the next is a score. In the next two columns the counters have been left as they lay on the table without gathering up the groups of five and replacing them by a single counter above the normal line.

The appearance of the dot-diagrams in the Exchequer record books is convincing evidence of the use of counters even after Arabic notation had been introduced; but there are other reasons for believing that the coming of Arabic figures did not for some time affect the method of reckoning. Not only were Nuremberg tokens still being manufactured in large numbers for use in various European countries, including England,[2] for some time after the general change from Roman to Arabic figures during the first half of the seventeenth century, but textbooks on Arabic arithmetic continued for some time to include instructions on the use of counters.

Mathematicians, of course, appreciated the advantages and greater efficiency of the Arabic system more readily than others, and no doubt they had been using it for computation long before it was understood and accepted by merchants and accountants. Many textbooks and treatises were written by mathematicians in the sixteenth and seventeenth centuries with the object of teaching the new science of 'Algorisme', which it was at first called; but some of them recognised that

Scholer. Then I perceiue Numeration: but I pray you how shall I doe in this art to adde two summes or more togither.

ADDITION.

Maifter.

The eafieſt waye in this art, is to ad but two ſums at once togither: howbeit, you may add moꝛe, as I wil tel you a, non. Therefoꝛe whē you wil add 2 ſums you ſhall firſt ſet doꝛn one of thē, it foꝛceth not which, and then by it dꝛaw a line croſſe the other lines. And afterward ſet doꝛn the other ſum, ſo that that line may be betwéen thē, as if you woulde add 2659 to 8342, you muſt ſet your ſums as you ſée here. And then if you liſt, you maye adde the one

to the other in the ſame place: oꝛ elſe you may ad them both togither in a new place: which way, becauſe it is moſt playneſt, I will ſhew you firſt.

Therefoꝛe will I beginne at the vnits, which in the firſt ſume is but 2 , and in the ſecond ſumme 9, that maketh 11. Thoſe do I take vp, and foꝛ them I ſette 11 in the new roome, thus.

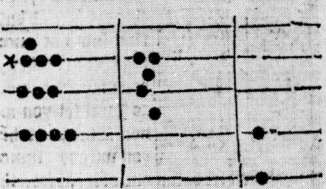

Then doe I take vp all the Articles vn, der a hundꝛed, which in the firſt ſumme are 40, and in the ſecond ſumme 50, that maketh 90 : oꝛ you may ſay better, that in the firſt ſumme there are 4 articles of 10, and in the ſeconde ſumme 5, whiche maketh 9, but then take héde that you ſet thē in their righ

---

the counter-board was still in common use, and they showed how the old method could still be used in combination with the new Arabic figures.

Among the best known textbooks in the English language were those of Robert Recorde. A chapter in his *Ground of Artes Teaching the Worke and Practice of Arithmetik*, first published in 1542, showed the use of the abacus method with Arabic figures. This chapter was retained in subsequent editions for more than a hundred years:

In dealing with addition the 'Maister' says to his pupil:

'The easiest way in this arte is to adde but two summes at ones together: how be it, you maye adde more, as I wil tel you anone. therefore whenne you wylle adde two summes, you shall fyrste set downe one of them, it forceth not which, and

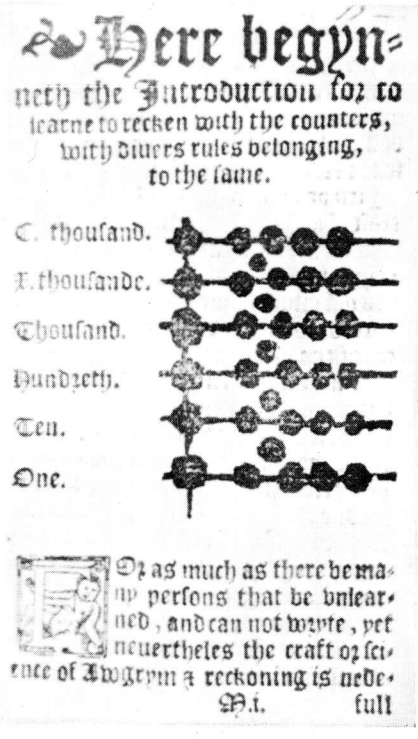

✍ Here begyn-
neth the Introduction for to
learne to recken with the counters,
with divers rules belonging,
to the same.

C. thousand.

X. thousande.

Thousand.

Hundreth.

Ten.

One.

For as much as there be many persons that be unlearned, and can not wryte, yet
neuertheles the craft or science of Awgrym & reckoning is nede-
M.i.   full

*'For as much as there be many persons that be unlearned, and can not wryte,
yet nevertheles the craft or science of Awgrym and reckoning is nedeful . . .'*

then by it draw a lyne crosse the other lynes, And afterwarde
sette doune the other summe, so that that lyne maye be
betweene them: as if you woulde adde 2659 to 8342, you must
set your sumes as you see here.

'And then if you lyst, you maye adde the one to the other in
the same place, or els you may adde them bothe togither in a
new place: . . .'

The diagrams in the book show how the counters are placed;
the method is described and explained in Chapter Five.

Recognition of the need to continue the old method for the
benefit of those who found 'Algorisme' difficult to understand,

or who had not yet learned to use pen and paper, was shown in *An Introduction to Algorisme* printed by John Awdeley in 1574, illustrated on the facing page.

The use of counters for reckoning was also described in books by Regius in 1543, *Utriusque Arithmeticas Epitome*, and Herbestus in 1577, *Arithmetica Linearis*. In France Le Gendre still found it necessary nearly two hundred years later, in 1753, to add a chapter to his *Arithmetique en sa Perfectionne* describing calculations 'par les jetons'.

Reading and writing, and the use of paper in general, were still the privilege of the few; and it should be remembered that such schools and universities as there were had no women students. It has been suggested that the needs of the distaff side in large establishments demanded the retention of the use of counters long after men had accepted pen-reckoning.

One of the most conservative institutions in this respect was the British Treasury. When the Norman kings began a regular system of taxes in the twelfth century the Exchequer was set up to deal with them. The name 'Exchequer' came from the fancied similarity of the table or cloth, with its lines or squares, to the newly introduced game of chess. From the twelfth century onwards the Exchequer Table was used for Treasury calculations until about the end of the eighteenth century. It was the custom throughout this time to record the amounts agreed between the taxpayer and the officials by cutting 'tally-sticks'. Presumably the amounts concerned were also entered in books, but the receipt for payment took the form of the 'foil' cut from the tally-stick until this system was abolished by the Statute of 1783. An indented paper 'cheque' was then given instead. The Exchequer Table probably went out of use at about the same time, but it was agreed that the old order should not be entirely abolished in the lifetimes of the two chamberlains then in office. The mass of tally 'stocks' held by the Treasury was ordered to be destroyed in 1834, but the fire got out of hand and the House of Commons was burnt down, a dramatic ending to the story.

There is abundant evidence of the use of the counter-casting

E

method from the twelfth century to the eighteenth in the large numbers of metal counters, or jettons, that are still to be seen in museums or in private collections, some of them with designs or inscriptions clearly indicating their purpose. One of the most interesting is a 'Nuremberg token' made by Hans Schultes and issued between 1550 and 1574. It shows on one side an abacus incorporated into a shield design and on the other a group of Arabic figures arranged into what appears to be a division sum, $891 \div 9 = 99$.

This 'rechen-pfennig' shows clearly that both methods were known and practised in the middle of the sixteenth century. The piece itself was made for use on a counting-board, but the manufacturer was anxious to show that he was aware of the new, Arabic, method of reckoning.

Unfortunately, however, nearly all the reckoning tables and cloths, which must have been very common two hundred years ago, have disappeared. Professor F. P. Barnard in his authoritative book *The Casting-Counter and the Counting-Board*, 1916, said 'the extreme rarity of specimens of the counting board is as remarkable, considering its general vogue in Western Europe during the six centuries from 1200 to the French Revolution, as is the survival of any examples of its more perishable substitute the reckoning cloth. . . . The (three) tables at Basle, the doubtful one at Nuremberg and the (five) reckoning cloths at Munich, are all that I have been able after considerable search and inquiry so far to discover in continental museums, while in this country there are none.'

'Table de payeurs, de la fin du XVIe siècle, confectionnée pour le receveur de l'Œuvre Notre-Dame en ces temps la'. Musées de la Ville de Strasbourg

There is, however, an excellent example of a reckoning table in a municipal museum at Strasbourg, and the 'Podington Shuffle-Board' at Hinwick Hall in Bedfordshire is possibly an English specimen.[3]

The 'shuffle-board' at Hinwick Hall, near Podington, is so-called because one of the end flaps has, roughly scratched upon it, lines and numbers, obviously used for a once fashionable game resembling our modern Shove-ha'penny. The dates 1634, 1718 (visible near the top left corner of the flap) and 1767 are cut into various parts of the table, which has been in the possession of the Orlebar family since the time of William Payne, whose initials appear on the flap (bottom right).

The initials W.O. and R.O. are attributed to William and Richard Orlebar who lived in the eighteenth century, but T.K. and R.H. were the initials of the butler (Thos. Knowlton) and the coachman (Robert Holloway) of a later period, cut, no doubt, after the table had been relegated to the servants' hall.

Attention has been focussed upon the shuffle-board flap, but the more serious use of the table, for reckoning, would be

The (probable) reckoning table at Hinwick Hall, near Podington. The positions of faint lines are shown by the cigarettes

shown by suitable lines, drawn or permanently incised upon the main surface. As it happens, it *is* possible to see faint traces of several such lines in the positions expected, distinguishable from random scratches by the fact that they are parallel to the sides of the table.

The lines have almost disappeared, which is not surprising after two hundred years of polishing and cleaning, and perhaps scraping from time to time. Even if there is doubt whether these were reckoning lines or not, the shuffle-board marked upon the end flap could well suggest a game played with the counters after a session of more serious occupation in casting accounts on the main part of the table.

Many reckoning tables, counter-boards and chequer-boards or cloths have been mentioned in wills and inventories (Barnard gives a list of 159 references), and it is most probable that there are others lying unrecognised in historic houses. They would best be identified by traces of straight lines on the surface

Shuffle-board game marked on the end flap

parallel to the edges, or perhaps by having a rim to prevent counters falling to the floor; but these features may have been removed when they were no longer needed and the tables were put to other uses.

The evidence of old documents and a long series of books on arithmetic, together with information about the few remaining reckoning tables and cloths and the large numbers of less perishable metal counters, all show that the change from counter-casting to pen-reckoning was, at first, slow and spasmodic. The period of most rapid change in England and Western Europe was the seventeenth century, and by the middle of the eighteenth century only a few reckoning tables remained in use.

Samuel Johnson, in his *Dictionary of the English Language*, published in 1755, used the present tense in describing 'a counter' as, among other meanings, 'A false piece of money used as a means of reckoning'; but his wording suggests that he may not have been familiar with genuine jettons, which were quite distinct from coins.

Bailey's *English Dictionary*, however, stated, in the 1740 edition, that a 'counter' was 'A Piece of Brass or other metal with a stamp on it, formerly used in counting, but now in Playing at Cards'. It also meant, he said, 'A Counting Board in a shop'. (We still use the word in that sense.) An 'abacus' was, according to Bailey, '(in Old Records) a Counting-table'. Clearly, the old custom was, by 1740, little more than a memory.

THE ABACUS METHOD

'Sitte doun and take countures rounde . . . And for vche a synne lay
thou down on (one) Til thou thi synnes have souzt up and founde.'
c. 1310 *(Oxford Dictionary)*

In his *View of the Origin, Nature and Use of Jettons*, published in
1769, Thomas Snelling said, 'A celebrated author is of the
opinion that the use of pieces to compute with was too natural
and simple not to be ancient, and probably was prior to Arithmetic itself; small stones, shells, kernels, etc., served those
nations for common use who either knew not, or despised,
gold and silver, and is still followed by many nations of
savages: he carried the invention as far back as the sons of
Noah, who to ease their memory, and settle among themselves
the increase of their flocks, made use of this sort of calculation,
the Egyptians knew no other method, which Josephus informs
us they learnt from Abraham. . . .'

Perhaps Snelling, or his unknown 'celebrated author',
carried his arguments too far, and he is certainly wrong about
the Egyptians, but reckoning with counters of one kind or
another is undoubtedly an ancient art, and may well have been
known by ancient nations long before the Greek and Roman
empires.

The 'simple and natural' method is to count with 'pieces' in
one place up to an agreed number. When the group is completed the pieces, or counters, are removed and one counter
placed in a second place. When an agreed total is reached in the
second place the same procedure is used, and so on for the
third and subsequent places. The same total will be used in all
places if numbers are being reckoned, ten being the usual
group number, but different numbers may be used in the groups
if money, weight or length, etc., are concerned.

A simple version of the method was used in French colonies

in Canada in the eighteenth century. Three boxes were used, one for 'deniers', one for 'sols' and one for 'livres'. A counter was thrown into the appropriate box for every denier, sol or livre. When twelve counters were reached in the 'denier box' they were removed and one put into the 'sol box', and so on. In this way the colonists could easily but slowly add up the amounts concerned in a money system they found strange and puzzling, even though they may have been unable to write.

This account is taken from P. N. Breton's *Guide Populaire Illustré des Monnaies et Medailles Canadiennes* of 1912. Whether it was used widely or not seems doubtful; but jetons were still being made in France for use in French colonies late in the eighteenth century. In any case, it illustrates the basic simplicity of the method, even with a comparatively complicated money system.

Lines, instead of boxes, have been used to mark the places from very early times. The Salamis abacus has eleven lines, although there is no direct evidence as to how they were used. The Darius Vase shows places marked by written Greek signs. Little is known of the plan used by the Romans, but they may have used vertical lines set out in this manner:

M	C	X	I

with counters representing D, L and V on the intermediate lines.

There are, however, four possible ways of setting out the abacus, two using vertical lines and two with horizontal lines. For example, the number 6594 could be represented by any of the systems shown at the top of the following page.

The commonest arrangement shown in old textbooks was the third, with units, tens, etc., on horizontal lines; but the fourth system appears on some exchequer cloths and tables in continental museums. The usual way of dealing with money, however, was to place the counters between lines in vertical columns, as in the second diagram, adding columns for shillings, pence, and perhaps farthings.

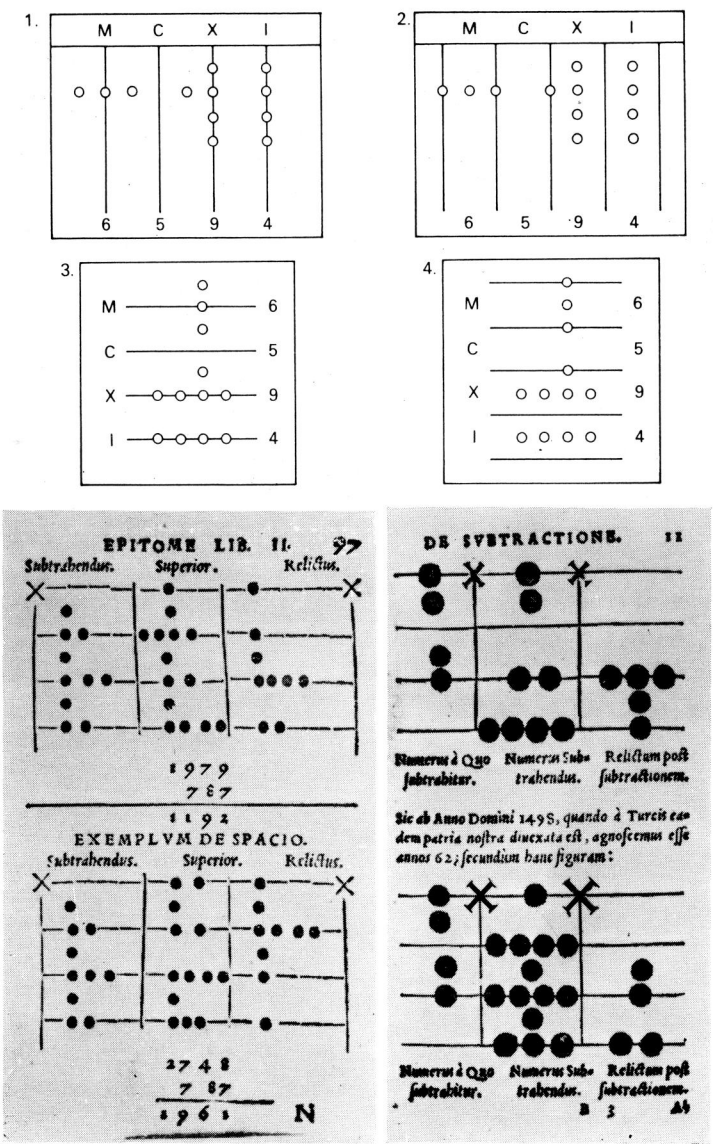

LEFT *A page from* Utrusque Arithmetica Epitome, *by H. Regius, 1543*
RIGHT *A page from* Arithmetica Linearis, *by B. Herbestus, 1577*

Early methods of reckoning have been fully described in books of instruction, such as those of Recorde, Awdeley, Regius and Herbestius. In all of these the units, tens, etc., are on the lines which are horizontal, with five, fifty and five hundred between them.

It was the usual custom in these times to mark the thousands line with a X (reminiscent of the crosses that appear on the Salamis abacus of about 500 B.C.—*see* p. 24). A comparable device is used today on the Russian schoty, which has a black bead on the thousands line, and sometimes another to mark the millions line.

An interesting variation was sometimes used, particularly in France, in which there were no actual lines on the table, the places being marked by a vertical row of jetons, the 'Arbre de Numeration'. This custom was followed by Le Gendre in 1753.

The following example of a subtraction sum is taken from

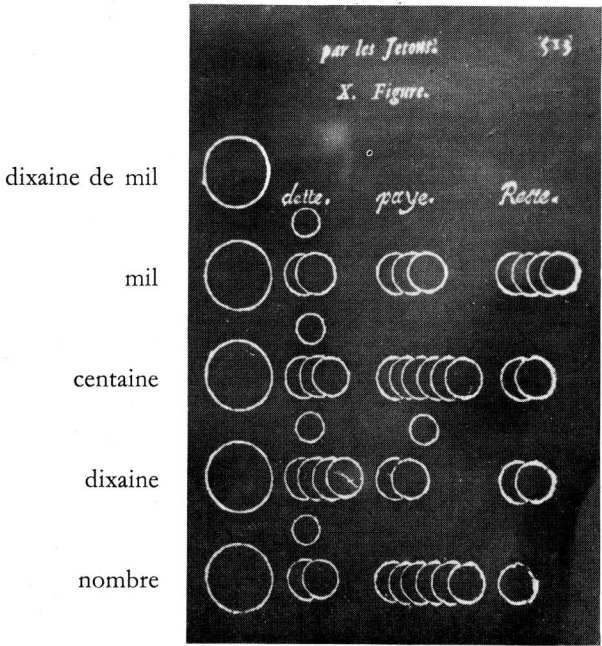

his book with the intriguing title *L'Arithmetique en sa Perfection*, published in 1753. The 'dette' shown in 7897. Of this 3676 is 'paye' leaving a remainder, 'Reste', of 4221.

These methods may look clumsy and slow to those of us who have become accustomed to Arabic figures and are proficient in using them for calculation, but when we consider that counter-casting was the normal method for dealing with accounts and calculations of all kinds for several hundred years it is evident that it must have been regarded as reliable and efficient.

A little practice with counters on a simple board will show that this is indeed true; and anyone who will take the trouble to work a few simple calculations may be surprised at the ease with which they can be carried out once he has become familiar with the feel of the counters. After all, the processes are very similar to those used by operators of modern Asiatic bead-frames, and recent experiments have shown that skilled manipulators of the Japanese soroban can work as fast as, and even surpass, experienced clerks using Arabic figures.[1]

Actual practice with counters is necessary, too, for a full appreciation of the size, shape and layout of a counting-board and of the reasons for the kind of moves that are made, including a number of short cuts that one soon begins to use. It is only by practice with counters that one realises why one should always push them from place to place and not attempt to pick them up; and one soon finds by experience what an advantage it is to have a low rim round the board.

When explaining the moves required for even a simple calculation detailed verbal or written descriptions tend, inevitably, to make the procedure seem longer and more complicated than it is in practice, and this is another reason why one should not rely upon merely reading about the abacus method, but should try it out for oneself. In doing so it will be found that every step can be followed by using Arabic figures on paper, but it is desirable that the temptation to think in terms of Arabic notation should be firmly resisted. One can, in fact, get a better idea of the older method by using Roman notation.

In the examples that follow, Arabic figures are used only for the convenience of the reader. It would be more realistic if instructions were given to the operator orally. When writing down the results it does not matter what notation is used.

Explanation of the methods used on the old counter-boards has been kept to a minimum in the hope that the reader will try to follow them with actual counters; but for those who prefer to try to imagine the moves, more detailed explanation is given in the appendix. There, the four examples shown in the following pages are worked out with the counters on vertical lines. In this way the abacus method can more easily be compared with the Arabic method, the columns M C X I being exactly comparable to the Arabic columns Th H T U.

Addition   The numbers are set down on the board side by side, and then the counters on each line are simply pushed together. A group of five counters on any one line is replaced by one counter in the space above.
Example:

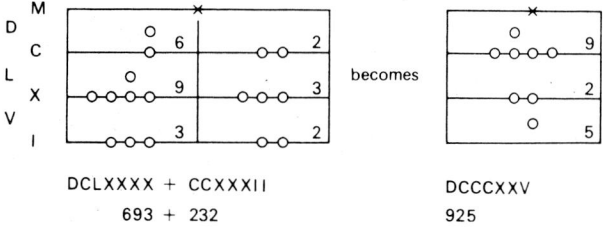

DCLXXXX + CCXXXII   DCCCXXV
693 + 232   925

As Robert Recorde says, 'Then doe I beginne with the smallest denomination'; i.e. one works upwards from the units line.

If several numbers are to be added they are dealt with in pairs.

Incidentally, the counter-board is very convenient for adding up long 'tots', building up the total as the numbers are called out. It works, in fact, just like an adding machine.

Subtraction   An example given by Robert Recorde will make the method clear:

To take 2892 from 8746.

First the numbers are set down, 'Then that I begin to Subtract the greatest numbers first (contrary to the use of the pen) that is the thousand in this example.'

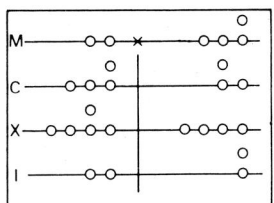

We could equally well begin with the units, but where there are blank lines this might require some such device as borrowing and paying back, which is avoided in using Recorde's procedure.

The counters to be subtracted, those on the left, are removed from the board as each line is completed. (This is different from the methods recommended by Regius, Herbestus and le Gendre in the examples shown on pp. 59 and 60.)

The steps followed by Recorde, then, are:

1 Take two counters from the M line.

2 To subtract eight hundred, take one from the M line and add two to the C line.

3 To subtract ninety, take one from the C line and add one to the X line.

4 Take away two units by removing the V counter and adding three.

The changes on the right-hand side of the board are:

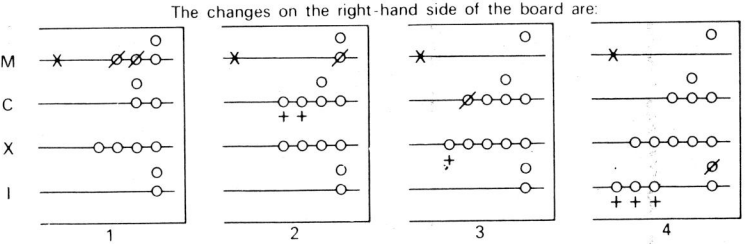

The answer can now be read as MMMMMDCCCLIIII or 5854.

Multiplication For multiplying, Recorde advises beginning with the highest denomination, but the order does not matter in this case. Multiplication by a single figure is a simple matter if multiplication tables are known. The number is set out to the left of the board, and as each line is dealt with, the counters on it are removed, the answer being shown on the right. Example:

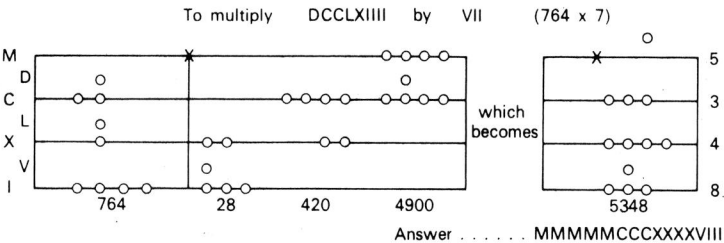

A practised calculator would build up the answer in its final form as he goes.

If tables are not known, the number can be set down the requisite number of times and the answer obtained by repeated addition. Factors could be used, of course.

'Long' multiplication is necessarily more complicated, but each step is a simple one.

Example: To multiply 634 by 523

The numbers are placed on the board, and '634' is then multiplied first by three, then by twenty and finally by five hundred.

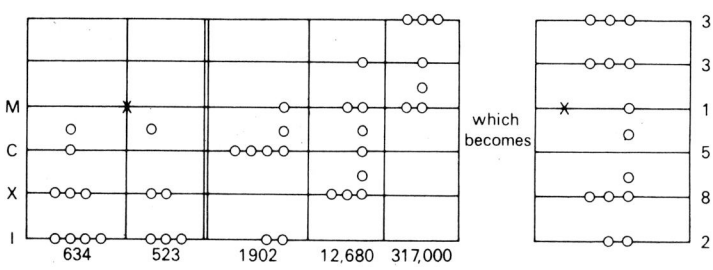

Answer 331,582

Sometimes the simplest method of multiplying on the counterboard is by doubling. For example, to multiply by seventeen one can double the number or sum of money four times and add the original number or amount.

Division Detailed descriptions of several methods used by various textbook writers are given by Barnard in his *Casting-counters and the Counting-board.* The methods used depended upon the nature of the problem and upon the sizes of the numbers involved.

Details of procedure in dealing with numbers vary but all writers agree in beginning with the highest orders of the dividend (as we do today in written arithmetic).

Some writers, including Robert Recorde, recommended dividing the board into three sections, setting down the divisor at the left-hand side. For example, in dividing 724 by 23 the board would look like this at the beginning of the operation:

	Divisor	Quotient	Dividend
M			
C			○ ○-○
X	○-○		○-○
I	○-○-○		○-○-○-○

Other writers suggested either setting the divisor down on the board *or* remembering it, and some asked the calculator to make a note of it.

In the above example the calculator begins with the C line where there are only 7 counters. He then moves to the X line on which there are 2 counters, making 72 X's in all. Three 23's can be taken out of 72. Therefore counters representing 69 are taken away leaving 3 on the X line.

	Divisor	Quotient	Dividend
M			
C			
X	○-○	○-○-○	○-○-○
I	○-○-○		○-○-○-○

34 units are now left and 23 can be taken out of this number once, leaving 11 as the remainder.

The final appearance of the board is then

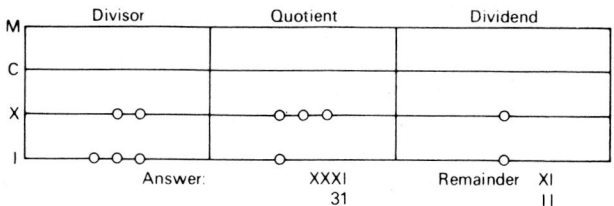

Division by a large divisor can be effected by repeated subtraction.

For money matters, both on the Continent and in Britain, pounds, shillings and pence, or some variation of these units, were in general use throughout the several centuries during which counter-casting was the reckoning method. In England farthings were added to the lower end of the scale if needed, while a space for scores of pounds was used instead of one for tens. The reason for this becomes apparent if one considers the practical use of the counters on the board. By this means one can treat the pounds in exactly the same way that the shillings are treated, and for all ordinary everyday transactions up to totals of not more than a few hundred pounds, this means a simplification of the operations.

The usual practice at the Treasury on the Exchequer Table, and no doubt in other large institutions, was to mark out the table, or cloth with vertical lines on the pattern already described in connection with the Tudor 'dot-diagrams' (*see* p. 40). This was described by Recorde as the 'Auditor's Use'. Such a system had been used at the Exchequer from the twelfth century and was described by Richard FitzNigel in his *Dialogues de Scaccario* written in the thirteenth century.

Plenty of room was needed (the Exchequer Table was ten feet by five) so that the counters could be placed accurately in position with no confusion.

Merchants seem to have used a rather simpler form of the same system for their smaller everyday amounts. The 'Merchants' Use' shown by Recorde had the values set out in this manner:

Scores of pounds	£100	£20	£20	£20	£20	
		£5				
Pounds	£10	£1	£1	£1	£1	
		5s				
Shillings	10s	1s	1s	1s	1s	
		6d				
Pence		1d	1d	1d	1d	1d

Recorde showed no lines but no doubt they were necessary.

The only counting tables or cloths still in existence on which the lines and markings are shown, all Continental ones, have a simple system of horizontal lines such as this (taken from a reckoning table at Basle):

M	
C	
X	
Lb	
s	
d	

The Strasbourg table (page 53), however, has an ingenious arrangement that can be used for either plain numbers (horizontal lines) or pounds, shillings, pence and farthings:

	Lib	s	de	f
C				
M				
C				
X				
I				

Nothing is known of any reckoning tables designed for length, weight or any other measures. No doubt these matters were dealt with without the need to have recourse to theoretical calculation.

It is evident from the various books, illustrations, reckoning tables and cloths and 'Rechen Meister' jettons that a variety of methods of reckoning with counters were used in medieval times, some of them being contemporaneous. All followed the same basic principles of what has been described here as 'the abacus method'. However, the rather elaborate variation of the counting-board known as the 'Exchequer Table', from its fancied similarity to the game of chess,[2] was used in dealing with royal accounts, both in England and in France as early as the twelfth century.

In the year 1110 Henry I addressed a writ to the barons of the exchequer ('baronibus de scaccario') for the collection of a tax of three shillings per hide for the marriage of his daughter. Apparently the 'scaccario' or 'exchequer' was already in being.[3]

A description of the scene in the Exchequer Chamber was given by Richard FitzNigel in the thirteenth century,[4] but the following picture comes from Hubert Hall's *Antiquities and Curiosities of the Exchequer*, written in 1891. A chapter on 'The Chess Game' describes the business of the Exchequer in full operation, with officials seated 'round what at first sight appeared to be a billiard table, with a dark cloth curiously patterned . . . the famous Exchequer Table.

'The central object of the chamber . . . was a table ten feet long by five in width, bordered by a ledge four inches high and covered with a dark russet cloth, divided into squares by intersecting lines, probably marked out with chalk, forming columns and spaces of account, within each of which a sum deposited had a certain numerical value according to its position towards the left hand of the reckoner; the column furthest to the right being for pence, the next shillings, the next pounds, and the remaining spaces scores, hundreds and

thousands of pounds respectively. . . . At about halfway down
the table's length sat the calculator. . . .

'. . . The chessmen of the Exchequer game were . . . counters
or dummy coins . . . of a size and appearance easily distinguishable from current money. For this purpose "besants" or the
depreciated "solidi" of the Eastern Empire were in requisition
at an early date. . . .

'And so the contest is duly waged . . . the metal counters . . .
being marshalled, advanced and swept off the board, just as the
pieces or pawns of the real game might have been played, till
the account . . . is concluded, and the mimic warfare terminates
in a truce between the parties for another six months at
least.'

Professor Barnard gave a similar account[5] based upon an
anonymous manuscript found with one of the Munich exchequer cloths. At the annual payment of accounts the Burgo-
master himself was in charge and was responsible for the
accuracy of the calculations. As he read out the accounts silver
jettons from a bowl were placed upon the cloth by a 'cavalier',
a man of noble or gentle birth, who would therefore be above
suspicion. A bishop was also present to see fair play.

As the various places became full, i.e. twelve pence, twenty
shillings, ten pounds (gulden) or ten gulden, the group of
jettons was removed and one jetton placed in the next higher
place.

As soon as the reckoning was finished, the 'gentleman' stated
the amount according to how the counters lay on the cloth. A
tally cutter might be present to record the amount, but entries
would probably also be made in an account book.

The 'counting-houses' in which the accounts of the larger
institutions were dealt with are sometimes mentioned in the
inscriptions on jettons specially made for use in them. The
sixteenth-century French jeton illustrated on p. 70, for example,
was used in the CAMERA COMPVTORVM REGIORVM,
the 'chambre des comptes' of the Royal Palace of Charles IX
(c. 1574).

The Spanish Netherlands jeton shown on p. 22 was made for

the CAMERA RATIONU INSULEN, which was the counting house for the town of Lille.

Jetons of silver used at the Royal Treasury of Louis XV of France are shown later (on pp. 83 and 84).

Silver counter of Mary, Queen of Scots, made about 1578 for use during her imprisonment in England. The inscription on the reverse VIRESCIT VVLNERE VIRTUS (Virtue flourishes in adversity) may be a protestation of innocence; and there may be some symbolic reference in the design to her three husbands, Francis who died early, Darnley who was murdered (cut down by the hand of God), and Bothwell who was still alive. A similar counter of Francis issued while he was still Dauphin of France, on which he described himself as SCOTOR REX (King of Scotland), bore on the reverse the legend CALCVLO ET RATIONE METIANDA OMNIA (All things are to be measured by calculation and reasoning). This strongly suggests that such pieces were, in fact, counters used for reckoning.

JETTONS

'Marchantes counters which nowe and then stande for hundreds and thousands, by and bye for odd halfpens or farthinges and otherwhiles for very nihilis.'

G. HARVEY, 1579

Methods of reckoning in common use in Western Europe from medieval times until the general acceptance of Arabic arithmetic in the seventeenth and eighteenth centuries are well attested by the evidence of numerous manuscripts and books, and by the survival of a few counter-boards and exchequer tables; but the most abundant, informative and concrete evidence lies in the existence of large numbers of the metal counters, often called 'jettons' that were actually used in the process.

The earliest jettons known to have been used in England were introduced from France in the thirteenth century, although some form of reckoning with counters must surely have survived from the Roman custom of using calculi and abacus throughout the Dark Ages and the Saxon period. Saxon communities had much building, trade and taxation, with well developed systems of coinage, and there must have been a good deal of reckoning; but there is hardly any evidence of how it could have been done.

Chaucer's reference, in the fourteenth century, to 'Augrim (i.e. algorithm) stones'[1] may be an indication that 'pebbles' were still being used even then.

According to Professor Barnard[2] 'the casting-counter of numismatic character had its origin in the royal house of France and . . . its use spread thence to the magnates of the land, to the public services, both central and local, to the people at large, and to the neighbouring countries'. Only one jetton known to Barnard was earlier than the reign of Louis XI (1226–1270). It is probable, however, that jettons were in use in Italy earlier

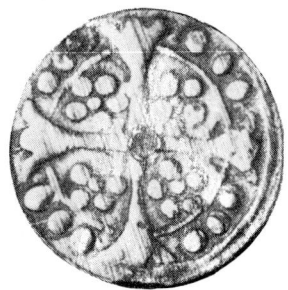

Early English jetton—possibly Edward II—early fourteenth century

than this; and there are records of the use of the Exchequer Table, which of course needed counters, in both England and France in the twelfth century.[3]

Jettons of the thirteenth and early fourteenth centuries were, according to Snelling, 'commonly called Black money'. Barnard described them as 'Anglo-Gallic' since it appeared that they were used in both England and France. Certainly, many jettons made at French mints circulated freely in England before the influx of Nuremberg tokens in the sixteenth century, but comparison of the earlier jettons with silver coins suggests that some, at least, came from English mints.[4]

There is probably a distinct class of Early English jetton, made, perhaps, from discarded coin dies.[5] Some jettons of this period have a small hole drilled into the centre of one side. Its purpose is unknown but it may have been intended to help to distinguish a jetton from a coin of the realm. (A similar mark appears on the silver counters made by Simon van de Pass for Charles I mentioned on pages 84–86.)

For the next two centuries all the jettons used in England were of French origin. They varied greatly but many of them show the fleur-de-lis motif and many bore religious emblems or legends.

The name 'jeton', usually spelt with two t's in English, comes from the French word 'jeter', to throw or cast. The phrase 'to cast accounts' comes from the use of jettons. They were not thrown on to the table, of course, but carefully pushed from

French—fifteenth century AVE MARIA GRACIA PLE(NA)

place to place, which, incidentally, explains why so many of them are well worn. The Dutch phrases 'leg-gelt' and 'leg-pfennig' come from the laying of the counters on the board; whilst the German phrase 'rechen-pfennig (or -pfening)' simply means reckoning-penny.

An early French name was 'méreau à compter'. In England 'compters' was sometimes used but it later became 'counters'. The delightful phrase 'A nest of cowntouris for the king' appeared in a Scottish Treasury account of 1496 whilst a document of 1540 values 'a nest of compters' at xviii s.

With no standardisation in spelling until much later there

French jeton—fifteenth century IETTES SEUREMENT IETTES

Abbey token—fifteenth century

were many variations in the early names. The French name appeared at various times as 'jetoir, jettoer, getoir, getor, gitour, jecton, gecton' as well as the more common 'gettes' and 'iettes'. (There was no distinction between I and J until later.)

The association of jettons with the keeping of accounts in large religious houses led to the use of the term 'abbey token';[6] and the name 'ward-robe counter' indicates that they were used in dealing with royal household accounts.

Struggles for power in France and the Low Countries in the fifteenth and sixteenth centuries were reflected in the appearance of jettons whose legends often expressed loyalty to one

French jeton—fifteenth century
Obverse: VIVE LE BON ROI DE FRANCE
Reverse: VOLGUE LA GALLEE DE FRANCE
(GALLEE may be a pun on Gaul, or Gallia)

Lombardy type of Nuremberg token

side or the other or showed the arms of a royal house.

From the early sixteenth century England seems to have depended entirely upon a small group of German manufacturers in Nuremberg. Nuremberg tokens were made in large quantities for use in England, France, the Low Countries and, of course, Germany. They often carried a head or bust of the appropriate monarch and sometimes bore a superficial resemblance to the current coinage although they were always made of brass. The names of the makers, Schultes, Lauffers, Krauwinckel, and a few others, usually appeared on one side or the other of these 'rechen pfennigs'. Nuremberg continued to be the main source of supply for the next two hundred years or so, as far as England was concerned.[7]

An interesting type of Nuremberg token showed the winged lion of St. Mark of Venice holding a Bible.

The earliest of these 'Lombardy' counters were introduced by merchants and financiers who came to England from Italy to replace the Jews after their expulsion at the end of the thirteenth century.

A common Nuremberg type showed on one side a design with three small open crowns alternating with fleurs-de-lis round a central rose; and on the other side a cross and orb appeared in a 'trilobe tressure', or fanciful shield. These counters often bore a proverb or a common saying such as HEUT ROT MORGEN TODT (Here today, gone tomorrow),

Common type of Nuremberg token
Obverse: MATHEUS LAUFER IN NURMBERG
Reverse: GOTES SEGEN MACHET REICH 1618
(From Proverbs X, 22. 'The blessing of the Lord, it maketh rich')

or DAS WORT GOTES BLEIBT EWICK (The Word of God endures for ever).

Later Nuremberg tokens varied greatly but some showed the heads of reigning English and French monarchs in crude imitation of silver or gold coins of the time. All English monarchs from Charles II to George II were represented, but

Obverse: Head of Charles II
CAROLVS.II DEI.GRATIA
Reverse: Shields of England, Ireland, Scotland and Wales
LAZA.GOTTL.LAVFF.RECH.PFE.COVNTER

Obverse: Head of James II
IACOBVS.II.D.G.ANG. SCO.FR.ET.HI.REX
Reverse: Royal coat of arms

there was often inconsistency and even gross carelessness. For example, reverses appropriate to Charles II still appeared with obverses of Anne and George II.

It will be noticed that most of these pieces are marked with the English word COVNTERS as well as an abbreviation of the German RECHEN PFENING.

Obverse: Heads of William and Mary
WILHELM.ET.MARIA.REX.ET.REGINA
Reverse: Shields of England, Ireland, Wales and Scotland
COVNTERS.IOHANN.WEIDINGERS.RECH.PFEN

Obverse: Head of Queen Anne
ANNA.DG.ANG.SC.FR.ET HIB.REGINA
Reverse: Sheilds of England, Ireland, Wales and Scotland
COVNTERS.IOHANN.IACOB.DIETZEL.RECHP.

Obverse: Head of George I
GEORGIUS M.BR.FR.H REX
Reverse: Royal arms
JOH CONRAD HOGERS.RECH.PF COUN.

Obverse: Head of George II GEORG.A D.G.M.B.F.REX
Reverse: Arms of England, Wales, Ireland, Scotland. (Stuart)
COVNTERS.IOHANN.IACOB.DIETZEL RECHP
(*N.B.: Another jetton with a similar obverse shows Queen Caroline*
on the reverse)

Although most jetons used in France were made there, and
were often of superior quality, brass tokens from Nuremberg
occasionally appeared, as these two examples show:

Obverse: Head of Louis XIV—1661–1715
LVD:XIIII.D.G. FR.ET NAV.REX.
Reverse: Royal arms CORNELI.LAVFFERS.RECHEN PFENING
1660–1676
(*Other Louis XIV tokens were made by Wolf and Conrad Lauffer*)

Obverse: Head of the Dauphin (of 1661? . . . Barnard)
GALLICUS DELIPHINUS
Reverse: A dolphin
WOLF LAUFFER RECH: PFENNIG MACH. IN NURNB:
J'AIME, ET SUIS AIMÉ
A play on the word 'dauphin' is also seen on earlier French jetons of the Dauphiné mint

Tokens produced at Nuremberg during the seventeenth and eighteenth centuries varied greatly in size and design, depending, apparently, on the whim of the designer, or perhaps the wishes of the customer. Mythical and historical figures were often shown, and occasionally a biblical scene. Some tokens were crudely mass-produced in thin brass, but others were finely executed in thicker metal.

Perhaps the most interesting to come from Nuremberg, and the most valuable to the student of the methods of reckoning, are those showing a merchant sitting at his counting-board. The earliest of the 'Rechen meister' tokens appeared about 1550. All show a table with a low rim and most of them show the abacus lines. These varied, perhaps because different methods were used or simply because the artist was not

attempting to be accurate. A printed alphabet appears on the reverse, presumably to popularise it among a largely illiterate population.

A bag or box, to contain money or a supply of jettons, is sometimes shown, and in one case an open account book lies on the table. There is, incidentally, a remarkable similarity between these illustrations and those of the Greek tax-collector and the Roman calculator shown on pages 26 and 27.

French jetons produced in the eighteenth century were often noteworthy for their high artistic quality, especially those made

for distinguished patrons. Some of them were of silver and occasionally a gold jeton was struck. When, for example, jetons were made for use in French colonies they might be of either silver or bronze, but a gold one might be presented to the king.

Jetons continued to be made in France up to the reign of Louis XVI, but the old method of reckoning ceased entirely with the Revolution.

In the reign of Louis XV jetons made in silver for use in dealing with the king's private accounts were marked TRESOR ROYAL.

Others, marked ORDINAIRES DES GUERRES, were used by permanent officials who dealt with the accounts of the

G

king's army. These were of silver; temporary officials serving for less than a year, 'extra-ordinaires', used jetons of brass or copper.

During the sixteenth, seventeenth and eighteenth centuries it was a custom in France for some important families and financial houses, following the royal example, to have specially designed jetons for private use. This kind of status symbol seems to have been rarely used in Britain, however, although a few early English, or Anglo-French, private counters are recorded in *Medallic Illustrations of British History*. John, Earl of Shrewsbury had his own silver counters inscribed with his name, IEHAN COTE DE SHROSBERI, his coat of arms and the family motto. In 1637 Bishop Juxon, who was both Bishop of London and controller of the Treasury, had counters referring to both these offices, GVIL LOND EPVS ET ANGLIAE THESAVR. *Medallic Illustrations* also shows counters of Mary, Queen of Scots; Francis II as King of Scotland; James VI of Scotland; the Earl of Leicester; and a doubtful one of Henry VIII and Katherine.

Mass-produced Nuremberg brass tokens were in general use about this time but during the reign of Charles I the Dutch artist Simon van de Pass, resident in England, was commissioned by the miniature painter Nicholas Hilliard to engrave a set of silver counters 'of the Royal Family', presumably for royal use. The event is mentioned in *Bryan's Dictionary of Painters* and the counters are recorded in *Medallic Illustrations*.

They were described in detail by Helen Farquar in the *Numismatic Chronicle* in 1916 and 1925.

The portraits include two of Prince Charles (afterwards Charles II) at different ages, which suggests that the counters were made over a period of years. Whether or not they actually passed into the possession of the Royal Family is not known, but some years later, in 1669, it was reported in the *London Gazette* that 'a Silver Box of Counters stamp't with Kings' and Queens' heads' was among the family plate stolen from the house of William Palmer of Grays in Essex by a band of seven men on horseback.

Some of these beautifully engraved counters are to be seen in the Department of British and Medieval Antiquities at the

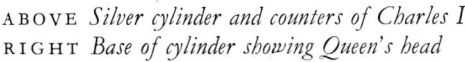

ABOVE *Silver cylinder and counters of Charles I*
RIGHT *Base of cylinder showing Queen's head*

British Museum and an excellent set is part of a private collection of jettons. They show portraits of all members of the family of Charles I from his grandmother St. Benissima Maria (Mary Queen of Scots) to his sons Charles and James and his nephew Charles Louis. His father, mother, wife, brother, sister and brother-in-law are all included. Other counters show the whole range of English monarchs from Edward the Confessor to Charles I.

With the counters is an exquisite silver cylinder large enough to hold about thirty-six counters which has a fine head of Charles on the lid and a poor one of his queen on the base.

Miss Farquar's account in the *Numismatic Chronicle* mentioned counters showing biblical scenes and L. A. Lawrence, in the same volume, referred to counters showing street cries. The same journal mentions an inventory in the Library of the Society of Antiquaries of the jewels of Henry VIII which includes several boxes containing counters, in sets of twenty-five or more, mostly of silver.

Although jettons are an interesting study in themselves, with intriguing designs and legends, and are sometimes beautifully made, they have attracted little attention up to the present. Numismatists have tended to ignore or despise them, perhaps because so many of the common varieties are badly made of thin brass with inaccurate, or even deliberately misleading, inscriptions. They may be worn by constant moving to and fro across the reckoning table, and, in any case, they have had little intrinsic value.

It was, of course, in the very nature and purpose of jettons that they should have no value in themselves, the value they represented depending entirely upon their position on the table. Any kind of counter, however poorly made, would serve as well. It was said in 1698 that in Muscovy 'Arithmetick' was performed 'by the help of Plumb-Stones instead of Compters'; and such small objects may have been used wherever there was a shortage of metal jettons.

The manufacturers of jettons were quite free from the severe conditions governing the minting and issue of coin of the realm and they took advantage of their opportunities. Coins were made only by officially approved mints to designs that were uniform throughout a whole country; but jettons could be made by anybody, and could even be bought from travelling pedlars. They were often made for individual business houses and designs could be varied at the wish of the patron or the whim of the maker. Consequently, they occur in great variety, and much can be learned from them about the customs and events of the times, the people for whom they were made and the ways in which they were used. Many jettons can be accurately identified and dated and can, therefore, be used to give valuable historical information.

Unfortunately, however, little reliable detailed information has been published about them, and there is, as yet, no comprehensive catalogue. Professor F. P. Barnard's scholarly book, *The Casting-counter and the Counting-board*, published more than fifty years ago, is the only authoritative book in the English language since Thomas Snelling's *View of the Origin, Nature and Use of Jettons* of 1769. Professor D. E. Smith published a monograph on *Computing Jettons* in America in 1921 but did not attempt to describe the jettons themselves. Professor Barnard restricted himself mainly to Continental jettons that were not already in his own large collection (now in the Ashmolean Museum); and jettons are mentioned only incidentally in the *Medallic Illustrations of the History of Great Britain*. We still have no full and illustrated account of those formerly in common use in this country.

On the continent none of the recognised reference books is recent. Neumann's *Beschreibung der Kupfermünzen* was published more than a hundred years ago and van Loon's *Histoire Métallique des Pays Bas* in 1732–7. Two French books are a little more recent, both appearing in 1904. They are Feuardent's *Jetons et Méreaux* and Florange's *Essai sur les Jetons et Médailles de Mines Français*. All are rare books not available to the ordinary collector.

Most museums have some jettons, for they often turn up among collections of coins or trading tokens, and there are very good classified collections of jettons in the British Museum and the Ashmolean; but for general information unless one is fortunate enough to have access to the rare 'Barnard' it is necessary to seek out and consult one of the few experts who have taken an interest in this subject.

At about the time that jettons were going out of use, card games in which counters were placed on the tables instead of the gold or silver coins were becoming popular in fashionable circles. Some of these were made of brass, for example the imitation 'spade guineas' of George III, and were similar in appearance to jettons. Gaming tokens are usually easily distinguishable by their inscriptions, but 'spiel-pfennigs', turned out in large quantities by the Nuremberg manufacturers when the demand for 'rechen-pfennigs' ceased, are sometimes so similar as to cause confusion.

Trading tokens, made of copper or other base metal to serve as small change before the official issue of regal copper coins, can often be identified by the name of the maker, trader or locality in which they were used, or by the portraits or other illustrations they carried. Advertising and commemorative tokens usually speak for themselves, but it is not always easy to distinguish French commemorative medalets from some of the well-made jetons that were fashionable in the later eighteenth century in France.

Trading tokens have received a good deal of attention, and other kinds of tokens and counters make an interesting study, but jettons are a distinct class. Their history shows them to be in direct line of descent from Roman calculi and Greek pessoi. Their appearance may sometimes cause confusion with coins and other varieties of tokens but they were made and used specifically for' reckoning on counting-boards and exchequer tables and were used exclusively for this purpose from the thirteenth to the seventeenth and eighteenth centuries.

THE ABACUS IN ARCHAEOLOGY

'A great nuisance these men who reckon with pebbles and crooked fingers.'

Letters of Alciphron (c. A.D. 180)[1]

Much of what has been said in earlier chapters about the origin of the word 'abacus', about methods of calculation used in Greece and Rome, and about the possible use of counters in more ancient times than those, may be relevant to some archaeological problems that are still under active discussion. It will be useful, therefore, to summarise the main conclusions:

[*i*] The word 'abacus' has been used very loosely in the past and has been given some meanings that are not justified by the evidence. Its original meaning may have been no more than 'a flat surface'. In classical times any kind of flat surface was described as an abacus, including, of course, one on which it was convenient to use pebbles for calculation.

There appears to be little justification for any suggestion that a sanded table was ever used for calculation with pebbles, although such a table may have been used, at one time, for drawing *geometrical* figures (*see* p. 18).

It is, strictly speaking, a misuse of the word to apply it to a bead-frame calculating device.

[*ii*] Some four or five examples are known of a Roman type of bead-frame 'abacus', but these must have been quite exceptional in Roman times. The three bead-frames now preserved in the British Museum, the Bibliothèque Nationale in Paris and the Museo Nazionale Romano are certainly of Roman origin. The devices still in general use in China, Japan, Russia and other parts of Asia are similar in principle but developed much later.

[*iii*] There is ample evidence that the normal everyday method of reckoning in Greece and Rome was with pebbles on a table

or tablet (the abacus), on which two or three parallel lines indicated the positions of the units, tens, hundreds, etc.

[*iv*] The abacus method of reckoning evolved from the earlier use of pebbles, or other counters, for simple counting. Counters of some kind may have been used from very early times, and the gradual evolution of the ideas of 'group numbers' and 'place value' may have taken many thousands of years.

[*v*] The use of numbers in practical situations, and some simple methods of reckoning, must have been well established before the emergence of any kind of written notation. The earliest systems of notation, including cuneiform, hieratic and Roman, were suitable and convenient for recording purposes, for which, no doubt, they were invented; but they were not suitable for the purpose of calculation. During the two thousand years or so during which Roman figures were in general use throughout the Western World there was no particular difficulty in the matter of calculation, for this was done by the simple and effective method of moving counters on the abacus, or 'counting-board' as it was called in later times.

The need for a written sign for zero did not arise before the introduction of the Arabic system.

[*vi*] Games in which counters are moved in spaces on a board have been played from very early times. It is possible, however, that some of the games could have begun as a relaxation from the more serious business of reckoning on the abacus. Some Roman counters were undoubtedly gaming-pieces; others could have been used for either purpose; but most calculi were probably used for calculation, as the name implies.

In recent years there has been some discussion, and no little controversy, as to whether or not ancient stone circles, with Stonehenge as the outstanding example, could have been designed on some mathematical basis, and used for recording, or predicting, movements of the sun, moon and planets. The absence of tangible evidence in the form of marks or signs, and our ignorance of the spoken language used in those times, mean that we can never have more than a sketchy idea of the ability to calculate of the men who built and used the stone circles. Two things are certain, however. In the first place, the size and complexity of the monuments mean that there must have been among them men of a high degree of intelligence and practical ability; and in the second, the absence of a written language or system of notation in no way precludes the possibility that they could measure, count and calculate.

Number relationships could have been memorised, or indicated by rows of pebbles, bundles of sticks or knots on a string, none of which would have left visible evidence that could be found today.

In those days the time factor could have been of little importance. It is only comparatively recently that we have made a virtue of speed in calculation. A modern digital computer deals with complex calculations at high speed; but many such calculations could, if desired, be broken down into a succession of simple steps, each of which could be worked out, slowly but surely, with no more apparatus than a few pebbles and a board. Even by counting only, aided by pebbles, it would have been possible, very slowly, to add, subtract, and multiply (perhaps by the method of doubling). Some easy division would have been possible, and various number series (e.g. square numbers) could have been discovered.

There is no evidence that this kind of arithmetic was known in Western Europe before the Roman period; it is mentioned here only to suggest what was possible without a written notation. Men of the earliest stone circle periods could have observed the recurring cycles of the sun, moon and planets using pebbles, or other counters, to mark the events as they occurred. As the piles or rows of pebbles grew, perhaps very slowly, the significance of recurring numbers could have been realised.

Some indication that this kind of primitive arithmetic might have occurred even at a very much earlier period appears in the 'notational sequences' examined by Alexander Marshack in an article on 'Lunar Notation on Upper Paleolithic Remains'.[2]

He speaks of 'thousands of notational sequences found on the "artistic" bones and stones of the Ice Age and the period following, as well as on the engraved and painted rock shelters and caves of Upper Paleolithic and Mesolithic Europe'. He considers in detail three examples, one an Azilian cave painting, another an engraved Upper Paleolithic mammoth tusk from the Ukraine and the third an Aurignacian engraved bone from Czechoslovakia. All show evidence that strongly suggests possible attempts by these very early people, living some thousands of years before the building of Stonehenge, to find some kind of numerical order in sequences of natural events.

Whether or not this kind of evidence can be accepted as an indication of the great antiquity of the first uses of numbers, the finding of flat round discs of stone, shale or clay on sites of all kinds from early Jericho[3] down to the Iron Age (including Stonehenge where a few blue-stone 'counters' were found) suggests the knowledge of ways of counting, and probably a certain amount of reckoning, long before the first written records were made.

We can be much more certain about methods of calculation in classical times for here the archaeological evidence is supported by literary references, or vice-versa. There are many references, from Herodotus onwards, in inscriptions and manuscripts to the use of pebbles on the abacus.[4] Weights, measures and money were in common use and inscriptions show that as early as the fifth century B.C. taxes were levied and interest was charged on money lent.[5] It is not surprising, therefore, to find that Greek abaci, of which at least twelve examples have been found, were relatively complex in form.[6] These are all stone or marble slabs, too heavy to be moved easily and with permanently engraved lines and characters. No doubt wooden boards or tables served for normal everyday use.

It may seem strange that although there are numerous Latin references to Roman abaci, and even to the teaching of 'pebbling' to Roman boys,[7] no example seems to be known of an undoubted Roman abacus found during excavations. However, this need cause no surprise because there is an exact

parallel in the dearth of counter-boards and exchequer tables which were in common use in England as late as the seventeenth century.[8] The explanation is not difficult to find. In the first place, any convenient flat surface would have served the purpose and specially designed tables would be rare; and in the second, a wooden board or table would be destroyed or become unrecognisable with the passage of time, or converted to some other use. Fortunately, a few illustrations showing Greek and Roman calculators in operation have survived to place the matter beyond doubt.[9]

There is evidence, too, from archaeological and historical sources in the New World. For example, the use of pebbles for counting and reckoning by primitive Indian tribes and of grains of maize threaded in rows of ten, as a form of abacus during the Maya civilisation. The detail above, from an Aztec vase found in Guatemala, shows tax being offered in the form of bags of cacao beans. The value is calculated by the seated man, using an abacus which appears to have a hook to hang it up by. The scene is remarkably similar to that shown on the Darius vase.

THE ABACUS IN EDUCATION

'Ambition, Distraction, Uglification and Derision.'
LEWIS CARROLL, *Alice in Wonderland*

Great progress has been made in recent years in devising new and better ways of introducing mathematical ideas and procedures to children in primary schools; and much new thought has been given by many teachers to the early stages in which children learn to count and, later, to reckon. It is, unfortunately, still true, however, that few teachers have a thorough understanding of the principles on which our number system is based or a clear idea of its history.

A simple test will show whether or not one understands the basic principles of Arabic notation. Imagine that letters of the alphabet are to be used instead of Arabic figures, with A standing for one, B for two, C for three, and so on up to I for nine. How would you them represent ten, eleven, twelve, and numbers up to, say, twenty; and how would you write one hundred?

What is often forgotten is that although the Arabic system has a decimal basis, only nine figures are used, with the zero sign inserted where necessary to keep them in their proper places. J is not needed, therefore, when letters are used. Ten becomes AO, eleven is AA, twelve AB, etc., up to BO for twenty. One hundred becomes AOO.

The history of common weights and measures, and of the peculiar abbreviations that are used for some of them, remains an intriguing mystery to many who are teaching children how to use them; and it is rare to find a teacher, or anyone else for that matter, who understands how calculations were made and accounts kept before the introduction of Arabic figures. Too often it is stated, even in authoritative books, that when

Roman figures were used there must have been difficulty because of the absence of a sign for zero. This would have been true only if people of the time had been foolish enough to try to do their 'sums' in the same manner that we now use Arabic figures, i.e. by writing on paper. Instead, it was the practice to write the amounts concerned, using cursive Roman numerals, to perform the actual calculation with counters on the abacus or counting-board, and to read off the result from the counters as they lay on the board. It could then be recorded on paper in the written notation.

When the counting-board began to be replaced, in the seventeenth century, by the new method of 'pen-reckoning' many people found the Arabic figures confusing especially the enigmatic sign for zero. At this time there were few schools, but, with the rapid development of a general system of education in the nineteenth century, the problem arose of how to teach Arabic arithmetic to young children.

Classes in the schools were large and there were few trained or qualified teachers. Inevitably, the practice arose of formal teaching of rules to be followed, with little or no attempt to explain them. It is only in recent years that teachers have finally thrown off the rigid, unenlightened, formal tradition and have endeavoured, instead, to use methods that try to ensure that the children understand what they are trying to do.

Much attention has been given to ways in which 'number concepts' develop in a child's mind, and one result of this has been the appearance of a variety of types of 'structural apparatus'. Properly used, they can undoubtedly be of great value, and many teachers have implicit trust in the particular type of apparatus they have chosen. There is, however, a fundamental objection to most of these devices. They give a different picture to each dimension, units, tens, hundreds, etc., whereas it is a basic principle of the Arabic system of notation that the *same* figures are used in each position. The noughts in such numbers as 10, 100, and 1000 are not intended to give a different appearance but to put the significant figure (in this case 1) in its proper place. They could be omitted if the position of the figure were known, for example if the numbers were written on squared paper.

This difficulty does not arise on the abacus where the same counters 'sometimes stand for more, sometimes for less'; indeed, it is difficult to see what advantages the newly invented types of apparatus have over the old method of counter-casting.

A historical approach, in which the experience of many thousands of years is telescoped into the short life of the primary school, would fall naturally into three essential stages:
[*i*] the use of counters to represent numbers,
[*ii*] introduction to the counter-board, and
[*iii*] the substitution of Arabic figures for the counters on the board.

It is true that counters have long been used in the early stages in infant schools, but anxiety of teachers, parents and sometimes the children to get on to the 'real' arithmetic of written figures has usually meant that counters have been abandoned too soon. Their greatest value lies in learning how to calculate with them.

In the earliest stages counters could be used to record numbers that may arise in purposeful 'play', for example in making a record of the number of cars passing the school in a certain time. A counter is added to a heap or row for each car that passes and as the number grows it is not difficult for the children to realise the value of arranging the counters (which may be pebbles, buttons or any small object) in groups of, say, five or ten. Larger numbers can then be managed more easily.

The next step, a vital one which must not be hurried, is to count in units up to ten (which they must accept as the basic number) over and over again in one place, the number of ten-groups being recorded by placing the same kind of counters in a second place. If a hundred is reached, i.e. ten groups of tens, a single counter is put down in a third place. This takes the children on to the simplest possible form of abacus.

Any total reached can be read off from the board in words. If a note is to be made of the total number this is best done by using Arabic figures as a form of shorthand, taking care that they all go into their proper places, using noughts to mark any empty columns.

It is not possible, here, to give more than a general indication of the kind of approach that could be used. It is to be taken for granted that it will include a wide variety of number experiences, with the orderly arrangement first of counters and then of written figures emerging as a central theme. Many number relationships and combinations will be encountered while still using the counters, including the parts of five and then of ten, simple additions and differences, and the easiest of the number series such as two, four, six, and three, six, nine.

The next stage comes when the need arises to add numbers of two figures or more, to find differences and to multiply. Division follows later. These are best done on the counter-board (*see* Chapter Five) until the methods are thoroughly understood; and even when the counters are superseded by Arabic figures it is a help to go back to the abacus in case of difficulty.

The word 'abacus' has been used here in its older meaning, i.e. a flat board or table with loose counters. A convenient size for school use would be about two feet by one and a half, with lines marking the spaces for units, tens, hundreds and thousands.

The lines could be either horizontal or vertical, but the latter corresponds more closely to the way that Arabic figures are used.

A set of about thirty counters, of any convenient kind, is sufficient for calculations up to four figures. Fewer counters are needed if the medieval custom is followed of using the lines for intermediate values, five, fifty and five hundred. In practice this is found to be a simplification, not a complication, for if there are more than four counters in any space it is difficult to see the number at a glance and they would have to be counted each time.

Space should be left below the working area for spare counters and there should be a rim round the board to keep the counters from sliding out of reach.

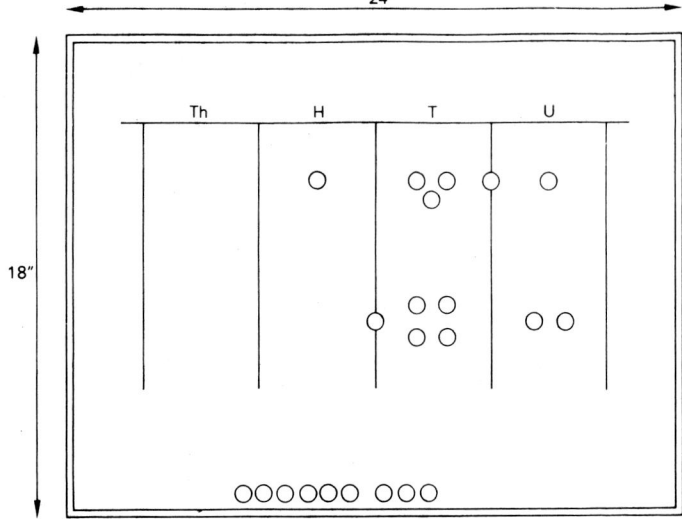

Counter-board showing 136 + 92

Counter-board showing £12.19s.11d.

When our money system ·is decimalised the same board could be used for money calculations. Meanwhile a simplified form of exchequer board could be used, although, by the time they are ready for it most children would be old enough to work with Arabic figures. Slower children, however, might find that the older method helped them to understand the newer one and it would interest all children to compare the exchequer board with a modern adding machine.

Teachers of arithmetic have always laid great emphasis upon accuracy in the use of written figures, but, in spite of their efforts, few children reach the high standard that is demanded in real life, especially where money is concerned. In this matter, the abacus has the advantage over the use of Arabic figures. Counters *must* be moved carefully and methodically into their proper places: precision is the very essence of the method, and the children will soon find that they cannot get on very well without knowing thoroughly all simple number combinations and tables.

A generation ago a 'counting-frame' was a familiar sight in every infants, classroom, and some of them, or smaller modern versions, are still often used for teaching children to count. Such devices are not without value, but to use them only for counting is to miss their real purpose and potentialities. A frame with three rows of ten beads may be used for counting up to thirty, but as an abacus it can be used for counting up to a thousand, as well as for a wide range of calculations.

Bead-frame computers, commonly described as abaci, are still used throughout China, Japan, Russia and other Eastern countries. The Japanese version, in particular, is a neatly made, sophisticated piece of apparatus which can be operated by experts with great speed and efficiency;[1] but it must be emphasised that skill in using the soroban is not attained easily. The instrument is not used in Japanese schools until the pupils reach their teens, and attendance at special abacus schools may be required.

H

Japanese soroban showing the numbers 123, 456, 789

For English children in primary schools a simpler form of bead-frame could be used. The Russian 'schoty', with ten beads on each wire, black ones being placed strategically, and with one wire for quarters (of a rouble), is a more suitable type to use as a model for an English version.

Russian schoty showing the number 123 (roubles)

Bead-frame computers have the advantage of being portable and easily stored, but as an introduction to the use of Arabic figures the old-fashioned counter-board, with its loose counters, is the better apparatus. It is more easily understood and can be used from the earliest stages. For some operations, including subtraction and division, the successive steps can be more closely related to Arabic arithmetic. It is only with handicapped children who cannot easily manage loose counters, especially the blind, that the bead-frame type of abacus is essential.

The abacus has suffered from the popular misconception that it is rather a crude device used by people in the past who could not manage their calculations because of the inadequacy of their written notation. A fuller understanding of the important part that it has played in the history of arithmetic may lead to a realisation of its value, even today, in the early stages of primary education.

Older children, especially those who are not fully confident in their 'pen-reckoning', might find that practice with the abacus would clarify their understanding of the principles of Arabic arithmetic, and there is much to interest them in the story of the use of calculi and jettons on the counter-board and Exchequer Table.

An addition sum in the Roman style. The answer has been found with pebbles on the board abacus, while another boy has found the same answer using a Russian schoty.

IX

CONCLUSION

When Thomas Snelling, quoting a celebrated but anonymous eighteenth-century author, published the view that 'The use of pieces to compute with was too natural and simple not to be ancient' he was expressing an opinion based upon his discovery of the ways in which jettons had been used during medieval and later times for reckoning, and upon the striking similarity between these methods and those used by the Romans with their abacus and calculi. In his time, the old method of counter-casting had almost disappeared, but his study of jettons, which were still found in abundance, led him to speculate upon the ways in which they could have been used; and he, and his un-known author, were shrewd enough to realise that the abacus method was more ancient that 'Arithmetic itself', and did not depend, therefore, upon the system of written notation.

There is abundant evidence of the ways in which counters have been used in Europe since about the eleventh century in the form of written descriptions of the scene and the method, the jettons themselves and a few reckoning tables and cloths; and although archaeological evidence about the use of the abacus in Greece and Rome is sometimes fragmentaty or circumstantial, there are sufficient literary references, with two or three illustrations, to show how easily everyday calculations were accomplished in classical times by the aid of a few 'pebbles'.

The finding of flat discs, which may have been used as counters, in excavations of sites as old as Jericho and Babylon, together with evidence of the playing of board-games that could have been derived from the abacus, suggests that the abacus method of reckoning may have been known in very early civilisations.

Throughout the very long period since the first discovery by primitive men of the method of counting in groups and the use of place value, the basic principles of the abacus method have not changed. Various systems of written notation have accompanied the use of the abacus, from time to time, and eventually the Hindu-Arabic notation brought in a new style of calculation in which the movement of counters disappeared and numbers were thought of entirely in the abstract.

The speed and efficiency of this 'pen-reckoning', once it was understood, swept away the older method, but over-emphasis on the written aspect of arithmetic tended to obscure the essential fact that the basic mathematical principles had not changed.

Now, in the second half of the twentieth century, we are turning once more to mechanical methods with calculating machines. Arabic notation is used at the beginning and end of the operation to record the amounts involved and to mark the keys to be pressed, but the result is achieved by the movements of the wheels.

In the earliest abacus pebbles were moved by hand; the latest form is an electrically operated machine.

APPENDIX

Pebbles used to build up a table

Finding square roots without calculation

Calculating with counters

Literary references to the use of counters

Classical references to the use of the abacus

Greek abaci

Makers of Nuremberg tokens

Bibliography

Notes

Pebbles, or counters, used to build up a table

e.g. of the squares of numbers. The numbers found can be recorded in any system of notation.

Pebbles					
1	4	9	16	25	36
Arabic					

Cuneiform

I	IV	IX	XVI	XXV	XXXVI

Roman

Finding square roots without calculation

The only apparatus needed is a stick to be used as a unit of length. It can be of any length but should be marked off in fractions, which can be found empirically. The longer it is, the greater the accuracy of the result that is possible.

A stick for measuring to two places of decimals can be made by marking off a hundred small units, e.g. barleycorns, and cutting off the remainder.

It must be assumed that Pythagoras' Theorem is known, viz. 'In any right angled triangle the square on the hypotenuse is equal to the sum of the squares on the other two sides.' (This was known to the ancient Egyptians.)

To find, for example, the square root of 13, mark off with the measuring stick two lengths in one direction and three units at right angles.

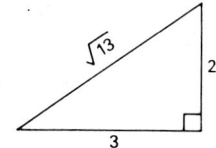

The hypotenuse will be the square root of two squared plus three squared, i.e. $\sqrt{4+9}=\sqrt{13}$. It will be found to be almost exactly 3 and 6/10 (actually 3·604).

The following sketch shows how the roots of the first few numbers could be found in this way.

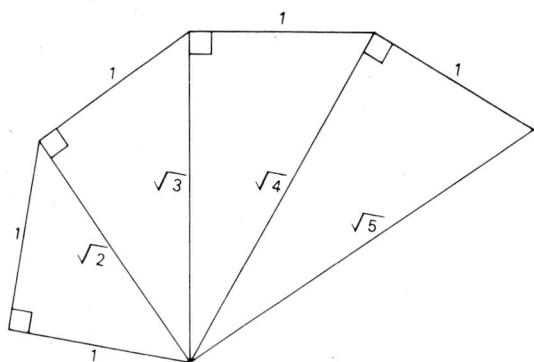

Calculating with counters

The examples shown on pp. 62–63 are worked out here in more detail; but instead of the horizontal lines favoured by Robert Recorde and other arithmeticians of the sixteenth and seventeenth centuries, vertical columns are used. The method is exactly the same, but in this form it is easier to compare the positions of the counters with the more familiar arrangement of Arabic figures.

ADDITION   E.g. 693 + 232

M	C	X	I
	o o o o	o	
		o	o
		o	o
		o	
	o	o	o
	o	o	o
		o	

Th	H	T	U
	6	9	3
	2	3	2

becomes

M	C	X	I
	o o o o	o	
	o	o	o
	o	o	o
		o	o
		o	o
		o	
		o	

or, rearranging the counters

M	C	X	I
	o o	o o	
	o	o	
	o		
	o		

Th	H	T	U
	9	2	5

N.B. With counters it is best to begin with the larger denominations.

M	C	X	I
O O O O		O O O	
O	O	O	
O		O	
		O	

Th	H	T	U
8	7	4	6

Take away two thousand, leaving

M	C	X	I
O O O O		O O O	
	O	O	
		O	
		O	

Th	H	T	U
6	7	4	6

Take away eight hundred, leaving

M	C	X	I
O	O O	O O O	
	O	O	
	O	O	
	O	O	

Th	H	T	U
5	9	4	6

Take away ninety (take a hundred, add ten), leaving

M	C	X	I
O	O O	O O O	
	O	O	
	O	O	
		O	
		O	

Th	H	T	U
5	8	5	6

Take away two, leaving

M	C	X	I
O	O O	O	O
	O	O	O
	O	O	O
		O	O
		O	

Th	H	T	U
5	8	5	4

Rearrange the counters

M	C	X	I
O	O O O		O
	O		O
	O		O
			O

M	C	X	I

A straightforward process if the tables are known. It was usual to start with the higher denominations.

M	C	X	I		Th	H	T	U
	o o o o		o					
	o		o			7	6	4
			o					
			o					

M	C	X	I			Th	H	T	U
		o o o		Seven x four				2	8
		o	o						
			o						
	o	o		Seven x sixty				4	2	
	o	o						
	o							
	o							
o o o				Seven x seven				
o	o			hundred		4	9	
o	o							
o	o							

M	C	X	I		Th	H	T	U
o	o	o o o		Total	5	3	4	8
	o	o	o					
	o	o	o					
		o						

Multiplication with larger numbers (long multiplication) was effected in stages. For example to multiply by 27 one could use factors, or multiply by 20 and then by 7, and add the results. Various other devices could be used, for example doubling. Thus one could multiply by 17 by doubling four times and adding the original number. To multiply by ten, all counters are moved one place higher; to multiply by five the result is then halved.

DIVISION E.g. 998 by 8

The procedures used depended upon the sizes of the numbers involved. With small divisors, and assuming that multiplication tables were known, division could be effected in a manner similar to the way in which it is now done with Arabic figures.

M	C	X	I

Begin with the highest denomination, i.e. with the hundreds.

∅ ∅ ○ ○ ○ ○
∅   ○   ○
∅   ○   ○
  ○   ○

(i) How many eights in nine (hundreds) ?
One.
Take away eight counters from the
C line, leaving: 198

∅ ∅ ○ ○
  ○   ○
∅   ○   ○
  ○

(ii) How many eights in the nineteen (tens) ?
Two.
Take away two eights, i.e. sixteen from
the X line, leaving: 38

  ○ ○
∅ ∅
∅ ∅
∅

(iii) How many eights in thirty eight ?
Four.
Take away four eights i.e. thirty two,
leaving a remainder of six.

Answer: One hundred and twenty four,
remainder six.

Division by a large number could be effected by repeated subtraction.

Literary references to the use of counters

SHAKESPEARE   (*Sixteenth/seventeenth centuries*)

The Winter's Tale, IV, 2.

Clown 'Let me see. Every 'leven wether tods; every tod yields pound and odd shilling; fifteen hundred shorn, what comes the wool to? . . . I cannot do't without counters.'

Othello, I, 1.

Iago '. . . a great arithmetician, one Michael Cassio . . . this counter-caster.'

Cymbeline, V, 4.

Gaoler '. . . pen, book and counters; so the acquittance follows.'

Julius Caesar, IV, 3.

Brutus '. . . to lock such rascal counters from his friends.'

Troilus and Cressida, II, 2.

Troilus 'Weigh you the worth and honour of a king,
        So great as our dread father, in a scale
        Of common ounces? Will you with counters sum
        The past proportion of his infinite?'

CHAUCER   (*Fourteenth century*)

Canterbury Tales. 'The Shipmannes Tale'.

    '. . . this marchant up ariseth,
    And on his nedes sadly him aviseth:
    And up into his countour hous goth he,
    To recken with himselven, wel may be,
    Of thilke yere, how that it with him stood,
    And how that he dispended had his good
    And if the he encresed were or non.
    His bookes and his bagges many on
    He layth beforn him on his counting board.'

'The Milleres Tale.'

    His alms geste, and bokes gret and smale,
    His astrelabe, longing for his art,
    His augrim stones, layen faire aparte
    On shelves couched at his beddes hed . . .'

HOBBES   (1651)

Leviathan, pt. i, ch. 4.

'Words are wise men's counters, they do but reckon with them, but they are the money of fools.'

Murray's *New English Dictionary*, 1893

c. 1310 *Know Thyself*, 38, in *E.E.P.* (1862) 131.
'Sitte doun and take countures rounde . . . And for vche a synne lay thou doun on Til thou thi synnes have souzt up and founde.'

1496 *Ld. Treas. Acc. Scot.* I, 300.
'A nest of cowntouris to the King.'

1515 Barclay, Egloges, iii (1570), c. ij/I.
'The kitchin clarke . . . Jengling his counters, chatting himselfe alone.'

1530 Palsgr. 648/1.
'I shall reken it syxe times by aulgorisme or you can cast it ones by counters.'

1579 G. Harvey, *Letter-bk.* (Camden), 66.
'Marchantes counters which nowe and then stande for hundreds and thousands, by and bye for odd halfpens or farthinges and otherwhiles for very nihilis.'

1540 *Act 32 Henry VIII*, c. 14.
'Item for euery nest of compters xviii s.'

Other references

1689 *London Gazette* No. 2498/4.
'A Robbery committed at the House of William Palmer, Esq. of Gray's in Essex by 7 men on Horseback . . . on Wednesday last the 16th. Instant . . . a very considerable quantity of Household Plate . . . a Silver Box of Counters stamp'd with Kings' and Queens' Heads . . .'

1698 Crull, *Muscovy*, 173.
'Arithmetick . . . which they perform by the help of Plumb-Stones instead of Compters.'

1673 Molière.
In the opening scene of Molière's *Malade Imaginaire* the hero is

represented as checking his doctor's bill with counters.

Classical references to the use of the abacus.

Herodotus (485–425 B.C.), in describing his travels in Egypt spoke of the Egyptians writing their figures and moving their pebbles from right to left, contrary to the Greek custom.

Herodotus (II, 36, 4) speaks of the Greek method of counting with pebbles in what must be vertical columns and so presumably a form of abacus.

Aischylos (*Agamemnon*, line 570) seems to refer to an abacus.

Aristophanes (444–385 B.C.) (*Vespae*, lines 332–3 and 656–7) suggests the physical form and use of the abacus. A character tells his father to do an easy sum 'not with pebbles, but with fingers'.

Aristotle (384–322 B.C.) mentions (Constitutions of the Athenians, 69), the stone slab used to count the votes in the courts.

Alexis (*Athenaios*, 117 c-e). A character in the comedy calls for an abacus and pebbles to do his accounts.

Demosthenes (385–322 B.C.) (XVIII, 229) refers to pebbling for calculations too difficult to be done in the head.

Lysias (fr. 50 Thalheim) mentions the abacus.

Diogenes Laertius (412–322 B.C.) attributes to Solon a remark that those who carry weight with tyrants are like the pebbles on an abacus because they sometimes stand for more, sometimes for less.

Polybios (V, 26) says: 'These men are really like the pebbles on reckoning-boards. For these, according to the pleasure of the reckoner, have the value now of a 1/8 obol (the lowest denomination on a Greek abacus) and at the next moment of a talent (6000 drachma) the highest denomination.'

Eustathius (Od. i. 107, p. 1397) speaks of an abacus as a tray covered by sand or dust used by mathematicians for drawing diagrams. He also speaks (Od. iv, 249) of 'the abacus on which they calculate.'

   Eustatius also speaks (Od. i, 107, p. 1396) of 'the abacus of Palamedes', a board for a game played with dice.

Persius (34–62 A.D.) (I, 131 f.) speaks of the sort of person 'who ridicules the numbers on the abacus and the partitions (metae) in its divided dust' . . . a reference either to an abacus marked out on the ground, or to a table abacus covered with sand.

Cicero (106–43 B.C.) and Juvenal (60–140 A.D.) both speak of aera (bronzes) as counters.

Cicero used the phrase 'ad calculos vocare aliquid', i.e. 'subject to strict reckoning', and spoke of counters as 'calculum subducere' (De Fin, ii, 18).

Juvenal (*Satire*, IX, 40) speaks of both the table and the counters:
'Computat . . . ponatur calculus, adsint
Cum tabula pueri; numera sestercia quinque
Omnibus in rebus, numerentur deinde labores.'

Pliny (23–69 A.D.) (*Hist. Nat.* XXXVI, 26, 67) mentions calculi and abaculos.

Capitolinus, speaking of the boyhood of Pertinax (126–193 A.D.) says: 'Puer litteris elementariis et calculo imbutus.'

Martial (II, 48) includes among his modest wants 'tabulamque calculosque'.

Horace (65–8 B.C.) speaks (Sat. I, vi, p. 75) of boys going to school with tablet and counters suspended from their left arm:
'Quo pueri magnis e centurionibus orti,
Laevo suspensi loculos tabulamque lacerto.'
(Loculos = 'little places' on the 'tabulamque' or 'tablet'. See caricature on p. 28.)

Alciphron. The *Letters of Alciphron* were written in Athens in the second century A.D. but purported to describe life in older times. In one letter (i, 26) a farmer writes of an unpleasant interview with a moneylender ending with the comment 'A nuisance these people who reckon with pebbles and crooked fingers'.

Greek abaci

Professor Mabel Lang of Bryn Mawr College, Pennsylvania, U.S.A., gives the following list of Greek abaci in a footnote to her article on 'Herodotus and the Abacus' in *Hesperia*, vol. XXVI, no. 3, 1957:

1   I.G., II, 2777. Pentelic marble. 1·49 m. by 0·754 m. 4·5–7·5 cms. thick. A set of 11 lines with centre line and crosses on third, sixth and ninth lines. A separate set of 5 lines. Numerals on three edges. This is the 'Salamis Abacus'.

2   I.G., II, 2778. Pentelic marble. 0·75 m. thickness 8·5 cms. Mutilated at left. No lines. A row of numerals.

3   I.G., II, 2781. Pentelic marble. 1·19 m. by 0·49 m. Thickness
    7·5 cms. Numerals along one edge. Six circles of various sizes.
4   'Aρχ.'Eφ., 1925–26, pp. 44–45, No. 156. White marble. 1·28 m.
    by 0·78 m. Thickness 8·5 cms. Sets of 11 lines and 5 lines.
    Crosses as on the Salamis abacus. One row of numerals.
5   Ibid., No. 157. White marble, right end lost. 0·80 m. by 0·64 m.
    Thickness including a rim 12 cms. 4 lines and a row of numerals.
6   Ibid. No. 158. White marble with rim. 1·305 m. by 0·645 m.
    Thickness 16·3 cm. 11 lines in centre and others in opposite
    corners. Uninscribed.
7   Ibid. No. 159. White marble fragment. 2 lines preserved.
    Uninscribed.
8   Ibid. No. 160. White marble fragment. 5 lines and 2 lines.
    Uninscribed.
9   I.G., XII, 7, 282 (Minoa). Two fragments of marble, broken
    on all sides; columns marked off by lines; numerals at top of
    each column.
10 I.G., XII, 5, 99 (Naxos). Fragment with row of numerals.
11 I.G., IX, 1,488 (Akarnania). Row of numerals.
12 B.S.A., XXVIII, 1926–27, pp. 144–45. Two edge fragments.
    Numerals along two opposite sides.

The 'casually converted roof tiles' mentioned by Professor Lang
(I.G., II², 2778, 2779, 2780) show numerical signs similar to those
of the Salamis abacus, but have no lines.

Makers of Nuremberg tokens

Georg Schultes	*c.* 1550–96
Hans Schultes	*c.* 1550–74
Damianus Krauwinckel	*c.* 1570
Egidius Krauwinckel	*c.* 1570–1600
Hans Krauwinckel	*c.* 1580–1610
Kilianus Koch	*c.* 1587–1600
Valentinus Maler	*c.* 1568–1611
Zacharius Jansen	*c.* 1575–1600
Hans Lauffer	*c.* 1607–45
Wolf Lauffer	*c.* 1618–60
Matheus Lauffer	*c.* 1618–25
Christian Maler	*c.* 1603–52

I

Contrad Lauffer	c. 1660
Cornelius Lauffer	c. 1660–76
Lazarus Gottlieb Lauffer	c. 1660–1700
Johann Weidinger	c. 1670–1700
Johann Jacob Dietzel	c. 1710–30
Johann Christian Reich	c. 1758–74

Others included the Hogers, the Lauers, and Jordan, of the eighteenth century.

The first eleven supplied a great proportion of the counters used in England.

Bibliography

Books describing reckoning with counters

WIDMAN, J., *Algorithmus Linealis* (*c.* 1488)

REISCH, G., *Margarita Philosophica* (1503)

CLICHTOVIUS, *Ars supputari tam per calculos que notas arithmeticas* (1507)

CLICHTOVIUS *De Mystica numerorum* (1513)

BLASIUS *Liber Arithmetice Practice* (1513)

CUSANUS, J. *Algorithmus Linealis Projectilium* (1514)

KOBEL, J. *Ain neu geordnet Rechenbeichlin auf den linien mit Rechenpfenningen* (1514)

SILICEUS, J. M. *Arithmetica* (1526)

RUDOLFF, C. *Kuntsliche rechnung mit der ziffer und mit den zalpfennige* (1526)

RECORDE, R. *The Grounde of Artes Teaching the worke and Practice of Arithmetike* (1542 and later editions)

REGIUS, H. *Utrusque Arithmeta Epitome* (1543)

BOCK *Rechenbuchlein auff der Linien und feder* (1544)

GÜFFERICH *Rechenbuchlin Auff der Linnen und Federn* (1546)

CATHALAN *Arithmetique . . . à chiffrer & compter par la plume & par les gestz* (1555)

ANONYMOUS *Arithmetique par les jects* (1559)

ANONYMOUS *Le livre des Getz* ('. . . la practique de bien scavoir conter aux getz comme la plume') (1563)

SPANLIN *Arithmetica* (1566)

PEREZ DE MOYA, J. *Tratado de Mathematicas* (1573)

AWDELEY, J. (Printer) *An Introduction of Algorisme· to learn to reckon wyth the Pen or wyth the Counters* (1574)

GORLA *Arithmetica* (1577)

HERBESTUS, B. *Arithmetica Linearis* (1577)

TRENCHANT, J. *L'Arithmetique de Jan Trenchant . . . Avec l'art de calculer aux Getons* (1578)

GUILLON, G. *Institution de l'Arithmetique avec les Gettons et la Croye* (1604)

HENISCHIUS, G. *De Numeratione Multplici* (1605)

LE GENDRE, F. *L'Arithmetique en sa Perfection* (1753)

SNELLING, T. *View of the Origin, Nature and Use of Jettons* (1769)

SMITH, D. E. *Rara Arithmetica* (1908)

SMITH, D. E. *Computing Jettons* (Monograph of the American Numismatic Society) 1921

SMITH, D. E. *History of Mathematics*, vol. II (Constable, reprinted 1958)

BARNARD, F. P. *The Casting-counter and the Counting-board* (Oxford University Press, 1916)

History of Mathematics

SMITH, D. E. *History of Mathematics*, vol. II (1925; reprinted, Constable, 1958)

SMITH, D. E. *Rara Mathematica* (Boston and London, 1908)

SMITH AND KARPINSKI *The Hindu-Arabic Numerals* (Ginn, 1911)

BALL, W. W. ROUSE *History of Mathematics* (Macmillan, 1919)

SCOTT, J. F. *A History of Mathematics* (Taylor and Francis, 1958)

CAJORI, F. *History of Mathematics* (Macmillan, 1919)

CAJORI, F. *History of Mathematical Notation*, vol. I (Court, Chicago, 1928–9)

YELDHAM, F. A. *The Story of Reckoning in the Middle Ages* (Harrap, 1926)

YELDHAM, F. A. *The Teaching of Arithmetic Through 400 Years, 1535–1935* (Harrap, 1936)

HILL, G. F. *The Development of Arabic Numerals in Europe* (Oxford University Press, 1915)

WRIGHT, G. C. NEILL *The Writing of Arabic Numerals* (University of London Press, 1952)

BOWRING, SIR JOHN *The Decimal System in Numbers· Coins and Accounts* (1872)

KAYE, G. R. *Indian Mathematics* (Thacker, Spink & Co., 1915)

STEELE, R. *The Earliest Arithmetics in English* (Oxford University Press, 1922)

History, general

WATERS, C. M. *An Ecomomic History of England* (Humphrey Milford, 1925)

BERRIMAN, A. E. *Historical Metrology* (Dent, 1953)

SINGER, HOLMYARD, ETC. *A History of Technology*, vol. II (Clarendon Press, 1954–8)

SALTZMAN, L. F. *English Life in the Middle Ages* (1926; reprinted 1960)

SALTZMAN, L. F. *English Trade in the Middle Ages* (Oxford University Press, 1931)

HASSALL, W. O. *They Saw it Happen* (Blackwell, 1957)

HECTOR, L. C. *The Handwriting of English Documents* (Arnold, 1958)

JENKINSON, H. *The Later Court Hands in England* (Cambridge University Press, 1927)

MURRAY, H. J. R. *A Short Story of Chess* (Oxford University Press, 1913, reprinted 1963)

MURRAY, H. J. R. *History of Board Games Other Than Chess* (Oxford University Press, 1952)

ORLEBAR, F. ST. J. *The Orlebar Chronicles* (Mitchell, Hughes & Clarke, 1930)

Archaeology

DARENBERG AND SAGLIO *Dictionnaire des antiquités grecques et romaines* (Paris, 1877)

PAULY-WISSOWA-KNOLL *Real-Encyclopädie der Klassichen Altertumswissenschaft* (Stuttgart, 1893)

FREEMAN, K. J. *The Schools of Hellas* (Macmillan, 1922)

HEATH, SIR T. *Greek Mathematics* (Oxford University Press, 1921)

MARROU, H. I. *A History of Education in Antiquity* (Sheed and Ward, 1956)

COWELL, F. R. *Everyday Life in Ancient Rome* (4th ed. 1966)

PEET, T. E. *The Rhind Mathematical Papyrus* (1923)

PEET, T. E. 'Mathematics in Ancient Egypt', *Bulletin of the John Rylands Library*, vol. 15 (1931)

NEUGEBAUER, O. *Exact Sciences in Antiquity* (Copenhagen, 1957)

SMITH, WAYTE AND MARINDIN *A Dictionary of Greek and Roman Antiquities* (1891)

LANG, M. Articles in *Hesperia*, XXVI, 1957; XXXIII, 1964 and XXXIV, 1965

BABELON, E. *Traité des Monnaies grecques et romains* Premiére Partie, 'Theorie et Doctrine', tome I (1901)

MARQUART, J. *La Vie Privée des Romains* (Paris, 1901)

PIHAN, A. P. *Exposé des signes de numeration usités chez les peuples anciens et modernes* (Paris, 1860)

FRIEDLEIN, A. P. *Die Zahl-zeichen und das elementaire Rechen der Griechen und Romer* (Erlangen, 1869)

MACALISTER, R. A. S. *Excavations of Gezer, 1902–1909* (1912)

KENYON, K. *Digging Up Jericho* (Benn, 1957)

MICHELL, H. *The Economics of Ancient Greece* (Cambridge University Press, 1963)

CARCOPINO, J. *Daily Life in Ancient Rome* (Heffer, Peregrine Books, 1964)

MEANEY, A. *Gazetteer of Early Anglo-Saxon Burial Sites* (Allen and Unwin, 1964)

Numismatics

CARSON, R. A. G. *Coins: Ancient, Medieval and Modern* (Hutchinson, 1962)

SEABY, P. *The Story of the English Coinage* (Seaby, 1952)

CHAMBERLAIN, G. C. *Guide to Numismatics* (English Universities Press, 1963)

PECK, C. W. *English Copper, Tin and Bronze Coins in the British Museum, 1558–1958* (London, 1960)

Textbooks on Arithmetic

LEYBOURNE *Cursus Mathematicus—Mathematical Sciences in Nine Books* (London, 1690)

TAYLOR, W. *The Arithmetician's Guide* (Birmingham, 1793)

WADDRINGTON, W. S. R. *Mechanical Arithmetic* (London, 1842)

PEACOCK, P. *Arithmetic* (about 1845)

Counter-casting and Jettons

BARNARD, F. P. *The Casting-counter and the Counting-board* (Oxford University Press, 1916)

SNELLING, T. *View of the Origin, Nature and Use of Jettons* (1796)

HAWKINS, E. *Medallic Illustrations of the History of Great Britain and Ireland to the death of George III*, compiled by E. Hawkins and ed. by A. W. Franks and H. A. Greuber, 2 vols. (London, 1885)

VAN LOON, G. *Histoire Métallique des Pays Bas* (Fr. ed. 1732–7)

NEUMANN *Beschreibung der Kupfermunzen* (1858–72)

FLORANGE, J. *Essai sur les Jetons et Medailles de Mines Français* (1904)

FEUARDENT *Jetons et Mereaux* (1904–7)

APLING, H. Article on 'The Casting-counter'. Seaby's *Coin and Medal Bulletin* (January 1962)

CHARLTON, J. E. *Standard Catalogue of Canadian Coins, Tokens and Paper Money* (1964)

BRETON, P. N. *Guide Populaire Illustré des Monnaie et Medailles Canadiennes* (1894 and 1912)

SMITH, D. E. *Computing Jettons* (Monograph of the American Numismatics Society, 1921)

LAWRENCE, L. A. 'On some early English Reckoning Counters', *Numismatic Chronicle*, Vth series, XVIII

KNOTT, C. G. 'The Abacus in its Historic and Scientific Aspects', reprinted and abridged from *Trans. of the Asiatic Society of Japan*, 1886, in Horsburgh, E. M., *Modern Instruments and Methods of Calculation* (London 1915)

The Exchequer

FITZNIGEL, R. *Dialogues de Scaccario* (The Course of the Exchequer), trans. by C. Johnson (Nelson, 1950)

POOLE, R. L. *The Exchequer in the Twelfth Century* (Clarendon Press, 1912)

POOLE, A. L. *From Domesday Book to Magna Carta*, 1087–1216 (Oxford University Press, reprinted 1958)

HALL, H. *Antiquities and Curiosities of the Exchequer* (1891)

TOUT, T. F. *Chapters in the Administrative History of England* (Manchester, 1920)

PALGRAVE, SIR R. H. I. *Dictionary of Political Economy* (1894)

Notes

INTRODUCTION

1 It is recorded that the average price of wool at Eton in 1572–82 was 20*s*. 9*d*. per tod of 28 lbs. Rogers, *History of Agriculture*.
2 'Jeton' is used for French counters, 'jetton' for English.
3 For discussion of the original meaning of the word 'abacus' see Chapter Two.

CHAPTER 1

1 R. L. Poole, *The Exchequer in the Twelfth Century*. Pierre-Louis Menon, writing in the bulletin of Le Club Français de la Medaille, 1964, said: 'En Angleterre, pays de tradition plus que le notre, cette manière d'etablir visuellement les comptes s'est maintenue à la Cour du Trésor (Echiquier) jusq'en 1826 . . .'.
2 'Awgrym' comes from 'algorism', an early Arabic name for arithmetic.
3 R. A. S. Macalister, *The Excavations at Gezer* (1902–5 and 1907–9) Vol. II pp. 229–30, and vol. III plate CCI.
4 A group of thirty sheep bones, labelled counters, can be seen in the Devizes Museum and there is another group in the Northampton Museum. Other instances of counters found on Saxon sites are given in the *Gazetteer of Early Anglo-Saxon Burial Sites* published in 1964.
5 K. Kenyon, *Digging Up Jericho*, pp. 57 and 58.
6 E. Mackay, 'A Sumerian Palace and the "A" Cemetery at Kish', part II, plate XLIV, no. 4. (*Field Museum of Natural History and Anthropology Memoirs*, vol. 1, no. 2.)
7 H. R. J. Murray, *A History of Board-games-other than Chess*.
8 e.g. the Podington Shuffle-board. *See* Chapter Four.
9 *See* appendix.
10 'For the Egyptians there could be no theory of numbers, only a practice.' T. E. Peet, *The Rhind Mathematical Papyrus*.
11 G. R. Kaye, *Indian Mathematics*.
12 M. Lang, *Hesperia*, XXVI, 1957; XXXIII, 1964; XXXIV, 1965.
13 K. J. Freeman, *The Schools of Hellas*, p. 104; and Smith, Wayte and Marindin, *A Dictionary of Greek and Roman Antiquities*.
14 K. Jackson, *Language and History in Early Britain*, p. 99: 'It

should be borne in mind that British was not a written language and that the only language of writing was Latin.'

15 T. G. E. Powell, *The Celts.*

16 ibid.

17 F. Cajori, *History of Mathematics*, vol. I, p. 26.

CHAPTER 2

1 C. J. Gadd, former Keeper of Egyptian and Assyrian Antiquities, British Museum, in private correspondence.

2 J. Levy, *Neuhebräisches und chaldäisches Wortebuch*, I 14a.

3 'Abacus is a table with the which schappes beth portayed and
   · ipeynt in powdre, and abacus is a craft of geometrie.'—1387 Trevisa, *Higden's Polychr.* (Rolls Scr.) VII 69 (Quoted in Murray's *Oxford Dictionary*). *See* also Liddell and Scott, *Greek-English Lexicon.*

4 Takashi Kojima, *The Japanese Abacus.*

5 W. Smith, *Dictionary of Greek and Roman Antiquities* (1842) mentions that 'By another variation the Abacus was adapted for playing with dice or counters. The Greeks had a tradition ascribing this contrivance to Palamedes, hence they called it "the Abacus of Palamedes".'

6 See list of Greek abaci, appendix, p. 114.

7 J. Carcopino, *Daily Life in Ancient Rome.*

CHAPTER 3

1 Quoted by W. W. Rouse Ball, *History of Mathematics.*

2 William Langland in a poem written in 1399 remonstrating with 'Richard the Redeles' (Richard II).

3 G. R. Kaye, *Indian Mathematics.*
   D. E. Smith, *History of Mathematics*, vol. II.
   Smith and Karpinski, *The Hindu-Arabic Numerals.*
   G. C. Neill Wright, *The Writing of Roman Numerals.*

CHAPTER 4

1 J. Palsgrave, Chaplain to Henry VIII and 'scholemaster to my Ladye Princes' (Mary, the king's sister).

2 The last Nuremberg counters for English use were issued in the reign of George II (1727–60). *See* p. 79.

3 The first suggestion that this might be a reckoning table was made by Mr. A. Turner of Bedford.

CHAPTER 5

1 In a book of instructions on *How to use the Japanese Abacus* the author, Kwa Tak Ming, recommends 'bead arithmetic' as taking only half the time needed by 'pen arithmetic' in fundamental operations; but agrees that the latter is more serviceable for intricate problems. *See* also p. 124.

2 H. R. J. Murray, *A Short History of Chess*. In the twelfth century chess was a fashionable game, newly introduced into England.

3 'Scacci' was used for chess in the twelfth century, and 'scaccarium' for either a chess-board or a reckoning board. 'Scheccarium' was a medieval spelling.—*Revised Medieval Latin Word List*, R. E. Latham.

4 R. FitzNigel, *Dialogues de Scaccario* (translated by C. Johnson).

5 F. P. Barnard, *The Casting-counter and the Counting-board*.

CHAPTER 6

1 Chaucer, 'The Milleres Tale'.

2 F. P. Barnard, *The Casting-counter and the Counting-board*.

3 A. L. Poole, *From Domesday Book to Magna Carta*: '. . . it (exchequer accounting) was based on the principle of the abacus which . . . was already known in the time of William Rufus (1087–1100).'

4 C. W. Peck, *English Copper, Tin and Bronze Coins in the British Museum*, p. 3.

5 L. A. Lawrence, 'Some Early English Reckoning Counters', *Numismatic Chronicle* (1938).

6 'Almost every abbey struck its own jettons or counters, which are thin pieces of copper, commonly impressed with a pious legend, and used in the casting up of accounts.'—Chatto, *Wood Engraving*, about 1880, p. 19.

7 Stock jettons were often bought from travelling pedlars. In 1550 an entry in the accounts of Wollaton Hall, Nottingham, by George Medley was 'For halfe a pounde of counters for my nece . . . to learne to caste with all—viij^d'.

8 Charlton, *Standard Catalogue of Canadian Coins, Tokens and Paper Money*, 1964.

CHAPTER 7

1 Although the *Letters of Alciphron* were written in Athens in the second century A.D. they were intended to appear as describing everyday life in the past. The letter i, 26 describes a farmer's visit to a moneylender. After an unpleasant interview the exasperated farmer remarks 'A great nuisance these people who reckon with pebbles and crooked fingers'. 'Crooked fingers' refers, presumably, to the practice of reckoning on the fingers (see p. 11); but 'pebbles' clearly refers to the abacus.

2 A. Marshack, *Science*, vol. 146, no. 3645, pp. 743–5 (November, 1964).

3 K. Kenyon, *Digging Up Jericho*, pp. 57–8.

4 *See* appendix for list of classical references.

5 *See* p. 11.

6 *See* appendix for list of Greek abaci.

7 *See* p. 28.

8 *See* p. 52.

9 *See* illustrations on pp. 26–29.

CHAPTER 8

1 Japanese abacus champions have beaten experienced operators of electric calculating machines in several contests, and it is still widely claimed in Japan that the abacus is superior to the machine (e.g. by Takashi Kojima in his books of instructions for the use of the abacus); but modern machines are being used more and more in industry and commerce, even in Japan, and calculating machines are now being manufactured in that country.

INDEX

abacus, Aztec 93
abacus, Greek, 23–6
abacus, origin of the word,
17, 89
abacus, Roman, 19, 20, 26–9
abacus method, 57 et seq.
abbey tokens, 74
Abraham, 57
Alexis, 16
Algorisme, 50
Algorismus, 36
algorithm, 71
Anne, Queen, 78
Arabic notation, origin of,
10, 35–6
Arabic notation replacing
Roman, 36 et seq.
Ashmolean Museum, 31, 88
astralagus bones, 8
augrim stones, 71
AVE MARIA, 73
Awdeley, John, 7, 51, 60
Awgrym, 50

Babylon, 102
Bailey's *English Dictionary*,
35, 56
Barnard, F. P., vi, 52, 65, 69,
71, 72, 87, 88
Basle, reckoning tables at, 52

besants, 69
Bibliothèque Nationale, 26–7
Black money, 72
board-games, 8
Boetius, 36
Breton, P. N., 58
Bristol, Mayor's Audits, 39–40
Bristol Record Society, 41
British Museum, 86, 89

calculator, Roman, 27
card-counters, 88
caricature of Roman school, 28
Carlyle, Thomas, 34
Caxton, 37
Celtic reckoning, 12
Charles I, 84–6
Charles II, 76
Charles IX of France, 69
Chaucer, 71
chess, 68
Chinese suan-pan, 20
Cicero, 16
Coligny, Bronze calendar, 12
compters, 73
counter-casting, 43 et seq.
counting- (or counter-) boards,
61, 93, 95
counting-houses, 69
cowntouris, 73

cumal, 12
cuneiform notation, 2–4

Darenberg and Saglio, 27
Darius Vase, 25–6
Dauphin, 80
Demosthenes, 16
deniers, 58
Dialogues de Scaccario, 66
Diogenes, 16
division, 65
dot diagram, Egyptian, 9–10
dot diagrams in Exchequer
    books, 43–8

Edward II, 72
Egyptian mathematics, 9
Exchequer, 43–8
Exchequer, origin of the
    word, 51
Exchequer cloths, 52
Exchequer Table, 1, 45, 66, 68
Extra-ordinaires des Guerres, 84
Eytocius, 12

Feuardent, 87
finger-signs, Greek, 11
FitzNigel, 66, 68
Florange, 87
Florence, 34
French colonies, 58

gaming-counters, 20
Gebert (Pope Sylvester II),
    x, 34
George I, 78
George II, 79
Gezer, 8
Greek calculation, example
    of, 11

Greek notation, 30–1
group numbers, 5

Henry I, 68
Herbestus, 51, 59
Herodotus, 11, 24
Hilliard, Nicholas, 84
Hindu notation, 11
Hinwick Hall, 53–4

INSVLEN, 22
Iron Age, 92

James II, 77
Japanese soroban, 20, 100
Jericho, 92, 102
jettons (jetons), 71 et seq.
Johnson, Samuel, 55
Jonson, Ben, 17
Josephus, 57

Kish, 8
Krauwinckel, 75

Lang, Professor Mabel, 11,
    23
Langland, W., 34
Lauffers, 75
Le Gendre, 51, 60, 63
leg-gelt, 73
leg-pfennig, 73
Leybourne, 38
Lille, 22
livres, 58
Lombardy, 75
Louis XI, 71
Louis XIV, 79
Louis XV, 83
Louis XVI, 83
ludi magister, 27

ludus calculorum, 21
Lullingstone, 20

Macalister, R. A. S., 8
Malta Library, Royal, 31
Marshack, A., 91
Mary, Queen of Scots, 70
medalets, 88
Merchants' Use, 67
méreau-à-compter, 73
multiplication, 64
Munich, 52
Muscovy, 86
Museo Nazionale Romano, 89

Neumann, 87
number, numeral, 14
Nuremberg tokens, 52, 75

Ordinaires des Guerres, 83

Palgrave, Sir R. H. I., 44
Persius, 17
pessoi, 25
place value, 6
plumb-stones, 86
Poddington Shuffle-Board, 53-5
Polybius, 16
Public Record Office, 43-8
Pythagorus, 36

rechen-pfennig, 73
reckoning-boards (-tables),
    23, 52
Recorde, Robert, 46, 49, 62
Regius, 59, 63
Reisch, 37
Roman numerals, origin of,
    31-2
Russian schoty, 20, 100

St. Andrews' Chapter, 39
Salamis abacus, 23-5
Saracens, 36
Saxon counters, 8
scaccario, 66, 68
schoty, 100
Schultes, 75
Scottish Treasury, 73
Senkereh tablet, 3
shuffle-board, 55
Snelling, Thomas, 57, 72, 87,
    102
solidi, 69
sols, 58
soroban, 20, 61, 100
spade guineas, imitation, 88
spiel-pfennigs, 88
Stonehenge, 90, 92

tally cutter, 69
tally sticks, 51
Taylor, W., 38
tod, ix
tokens (various), 88
Treasury, 51, 66
trilobe tressure, 75
Tyrol, gulden-groschen of, 37

Van de Pass, 84
Van Loon, 87
Venice, 75

Waddrington, W. S. R., 1
ward-robe counters, 74
wethers, ix
William and Mary, 77
Windsor Castle Accounts, 43-4
winged lion of St. Mark, 75

zero, 35-6